Y0-EHG-261

SIMPLIFYING
POWER SUPPLY
TECHNOLOGY

SIMPLIFYING POWER SUPPLY *TECHNOLOGY*

by Rajesh J. Shah

AC Delco Systems
Division of General Motors

An Imprint of
Howard W. Sams & Company
Indianapolis, Indiana

FIRST EDITION—1995

PROMPT® Publications is an imprint of Howard W. Sams & Company, 2647 Waterfront Parkway, East Drive, Suite 300, Indianapolis, IN 46214-2041

International Standard Book Number: 0-7906-1062-0

Library of Congress Catalog Card Number: 95-67558

Acquisitions Editor: Candace M. Drake
Editor: Natalie F. Houck
Assistant Editors: Rebecca A. Hartford, Karen Mittelstadt
Illustrators: Doug Cobb, Jyoti Shah
Typesetter: Leah Marckel
Cover Design by: Phil Velikan

Printed in the United States of America

9 8 7 6 5 4 3 2 1

To my father,
Jayantilal B. Shah.
Thank you.

TABLE OF CONTENTS

ABOUT THE AUTHOR

Rajesh J. Shah is currently working for AC Delco Systems on the General Motors electric car program. He has twelve years of experience in the field of power conversion. He has worked for Lambda Electronics, Branson Ultrasonics, Magnetek, and Valmont Electric. He holds a master's degree in Electrical Engineering from Polytechnic University, Brooklyn, New York.

PREFACE

This book simplifies the concepts of power supply technology to a level that a novice can understand; however, it assumes that the reader is familiar with basic electronic components and basic mathematics.

In general, universities and academic institutes do not offer as many courses in power supplies as they do in other areas of electronics. Due to the lack of exposure to the power supply systems, many feel that they do not have enough background to enter the field. On the contrary, anyone with some level of exposure to electronics can enter this field. The design of power supply systems is often described as black magic, but it is far from it. In fact, if understood properly, one would agree that it is an exact science. Although a tremendous amount of work is yet to be done in this field, it has come a long way in recent years. Many researchers have dedicated their careers to advancing the field of power electronics and a good amount of effort has been put forward in making contributions to the field. One could only appreciate the beauty of this field by first understanding the basic concepts such as the ones introduced in this book.

This book acts as an entry point into the field of power supplies. However, it can be used by experienced engineers to understand the basic concepts of various topologies. For them, this book will be a useful reference.

While there are those who may feel that power supplies are black boxes with a few terminals, and there is nothing to them, these people will find that there is more to the "black boxes" than meets the eye. I encourage them to spend a few seconds and flip through the pages of this book to see how much is involved in making the "black boxes."

This book is also written for engineering associates who, on a day-to-day basis, are bombarded by complex terminology and need basic explanations. This book provides simplified, yet detailed, operations of various topologies. With this basic understanding, associates can work effectively with experienced engineers and contribute to their design efforts.

Marketing and sales people, who are often put in situations where they have to present products to a highly technical group of people, need to know a little more about the technical aspects of their products than the general product catalog describes. This book will familiarize them with the technical terms that experienced engineers may use, facilitating their marketing efforts.

Last but not least, this book can be used by students who wish to pursue a career in the power supply industry or in academic research.

I would like to take this opportunity to thank my wife, Jyoti, for helping me prepare the manuscript and for spending long hours hand-drawing the illustrations. I appreciate her hard work and efforts to help me perfect the manuscript. I would also like to thank my six-year-old son, Neil, for constantly reminding me of the deadline by counting the number of days he had to play without me, and my two-year-old daughter, Sonia, who often helped me by turning the computer ON and OFF. I would like to extend special thanks to my brothers, Amit and Rupal, and to my sister, Purna. Without their tremendous support and faith in me, my career in the field of power supply would not have been possible. I cannot thank them enough. And to my uncle, Bhagwatilal B. Shah, and Ben Indumati J. Shah, thank you for being there when we needed you the most.

This book was written for readers from various backgrounds. With the basic understanding acquired from this book, the reader will be ready to dive into more advanced books. Go for it!

CHAPTER 1
AN OVERVIEW

The purpose of this book is to explain the concepts of power supply technology in detail, using a simplified approach. The power supply system consists of various sections, and we will examine these sections in detail. We will look at the input section, the regulators, the converters, and the control section. At the end, we will look at the specifications of power supplies and their applications. Following is a brief overview of the book.

Chapter 2 covers the basic concepts of various types of input circuitry, and numerous theories of operation are explained.

Chapter 3 covers the regulator circuits. This includes the commonly used topologies of linear regulators and switching regulators. The switching regulator section examines the Buck regulator, Boost regulator, Buck-Boost regulator, and C'uk regulators (switching regulators are examined in the ON and OFF states). Many books cover these important concepts in just a few lines, but this book covers them in detail. Basic equations are included to summarize the concepts.

Chapter 4 introduces a new set of topologies. These circuits are built upon the concepts of regulators covered in Chapter 3. Basic equations are also included to simplify the concepts. This chapter covers converter topologies: forward, flyback, push-pull, half-bridge, and full-bridge. It also explains the limitations of PWM technique at higher frequencies.

Chapter 5 penetrates complex mathematical jargon and brings concepts to a level that an entry-level person can understand. Due to the mathematical nature of this subject, some concepts are explained using simple mathematical derivations.

Chapter 6 introduces the reader to simplified explanations of the specifications of power supply systems. This chapter then introduces application-related configurations of the power supply systems.

The Glossary contains the definitions of words that are usually encountered when one is dealing with power supply systems.

CHAPTER 2
INPUT

2.1 EMI FILTERS

The main purpose of *electromagnetic interference* (EMI) filters is to reduce emissions, or noise. There are two types of noise that are of interest to us: conducted noise and radiated noise.

Conducted noise is so called because it is noise generated by the systems (such as power supplies) that is conducted to the input lines. This is undesirable because it goes back in the utility lines and creates noisy power lines, which can be a problem for other systems connected to the same power lines.

Radiated noise is so called because it is noise generated by the systems that is radiated in the air. Radiated noise can also be called transmitted noise. When the noise is transmitted in the free air, it can interfere with other systems.

Since it is not possible to have a completely noise-free system, designers try to reduce the noise levels of their systems. Usually, designers follow the guidelines set by some regulating agencies, such as the Federal Communications Commission (FCC) and the Verband Deutscher Electrotechniker (VDE). The amount of noise reduction is specified in these guidelines, and EMI filters have to be designed to reduce the noise per these guidelines.

2.2 RECTIFIER AND FILTER

2.2.1 RECTIFIER

Electrical power is usually generated and distributed in the form of *alternating current* (AC). Since many electrical systems require *direct current* (DC) as the input, it is necessary to convert the AC supplied by the utility companies to DC. DC is accomplished by first rectifying and filtering. The rectifying of AC is done by the rectifier section, and the main component used in this section is called a rectifier. A rectifier is a sub-class of the diode family, and the type of diode used in this section is normally known as a rectifier diode. The diodes can be arranged in a manner so that either a half-wave rectification or a full-wave rectification is accomplished.

The AC power can be either of single phase or multiple (poly) phase type. For a low power system, a single phase input is normally used, while a three phase input is normally used for a higher power system. Either single or three phase inputs are used for a medium power system; however, three phase inputs can also be used for a low power system.

In some electrical systems, the rectifier section is located after the line transformer, which is normally used for step up/step down and/or isolation purposes; but in modern switching power supplies, such a transformer is not desirable because one of the requirements of the switching power supply is to be light in weight and small in size. The presence of this transformer makes it very difficult for designers to meet the weight/size requirements because line transformers are bulky (and also costly). One way around using the transformer is to connect the rectifier section directly across the input line. Most switchers are designed in this manner, and they are known as off-line switching power supplies, or off-line switchers.

Commonly used configurations for the rectifiers are as follows:

Single phase half-wave.
Single phase full-wave center tapped.
Single phase full-wave bridge.

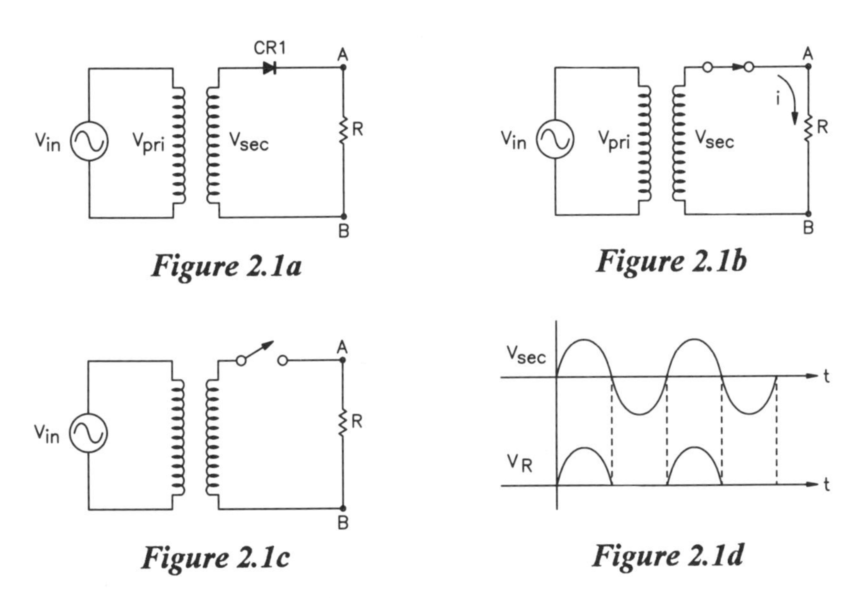

Figure 2.1a *Figure 2.1b*

Figure 2.1c *Figure 2.1d*

Single Phase Half-Wave

This configuration converts a single phase complete sine wave voltage into a single phase half-wave voltage. The direction of the rectifier diode determines the polarity of the output waveform, and the waveform could be either positive or negative.

For the diode arrangement in *Figure 2.1a*, only the positive waveform appears across the resistor (R).

When the AC voltage, V_{sec}, is positive as in *Figure 2.1a*, the diode (CR1) is forward-biased and will start conducting. (The equivalent circuit is shown in *Figure 2.1b*.) The current, *i*, will flow in the circuit and the voltage across R will be developed by the relation $v = i * R$. When V_{sec} is negative, then CR1 is reversed-biased, the circuit is open, and there is no current flowing in the circuit; therefore, the voltage across R is 0. (The equivalent circuit is shown in *Figure 2.1c*.) The output voltage will be zero for the entire negative portion of the V_{sec} cycle; therefore, the circuit will produce a positive waveform across R, as shown in *Figure 2.1d*.

For the diode arrangement in *Figure 2.2a*, only the negative portion of the cycle will pass, and the positive portion will be blocked; therefore, the circuit in *Figure 2.2a* will produce a negative waveform across R, as shown in *Figure 2.2d*. The reader may follow the above reasoning in deriving the negative waveforms. (See also *Figures 2.2b* and *2.2c*.)

Single Phase Full-Wave

In the full-wave configuration, the positive and negative portions of the V_{sec} cycle are passed on to the output. There are two types of full-wave configurations, namely center tapped and bridge.

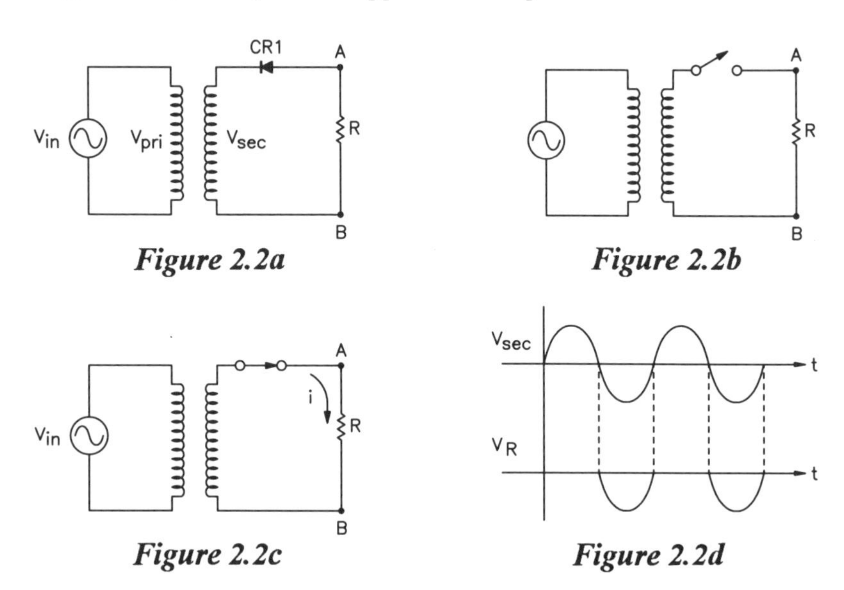

Figure 2.2a *Figure 2.2b* *Figure 2.2c* *Figure 2.2d*

Center Tapped

In this configuration, the secondary of the transformer is center tapped, meaning the secondary coil is equally divided in half. Each half of the coil produces voltages of both polarities, positive and negative. For example, V_{s1} with respect to the center tap is positive, and V_{s2} with respect to the center tap is negative.

During the positive portion of the cycle in *Figure 2.3a*, terminal 1 is positive with respect to terminal 2, and terminal 2 is positive with respect to terminal 3; therefore, terminal 1 is positive with respect to terminal 3. Another way of saying it is that terminal 3 is more negative than terminal 2 and terminal 1; therefore, the diode (CR2) connected to terminal 3 is reversed-biased and the bottom section does not conduct. Only the top section conducts because the diode (CR1) connected to terminal 1 is forward-biased. (The equivalent circuit for the positive portion of the cycle is shown in *Figure 2.3b*.)

The current in R flows from terminal A to B, developing a positive half-waveform (or one can say that the circuit is passing the positive portion of the sine wave).

During the negative portion of the cycle, terminal 1 is negative with respect to terminal 2, and terminal 2 is more negative with respect to terminal 3; therefore, terminal 3 is positive with respect to terminal 2 and terminal 1. The bottom section conducts because CR2 is forward-biased, and the top section does not because CR1 is reverse-biased. (The equivalent circuit is shown in *Figure 2.3c*.)

The current still flows from terminal A to terminal B, and the waveform that appears across R is positive; therefore, there are two positive half-waves. In other words, the positive portion of the input wave is duplicated, while the negative portion is "flipped." (See *Figure 2.3d*.)

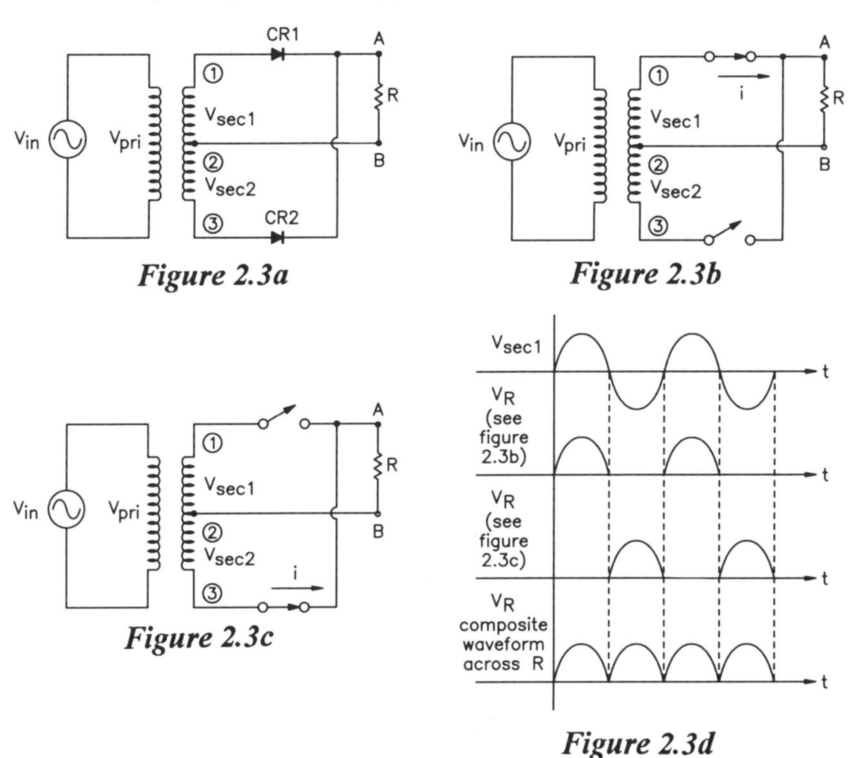

Figure 2.3a

Figure 2.3b

Figure 2.3c

Figure 2.3d

If the diodes were configured, as shown in *Figure 2.4a*, then the output would be a negative-going waveform. The reader may follow the above reasoning in deriving the negative-going output waveform. (See also *Figures 2.4b*, *2.4c*, and *2.4d.*)

Bridge

The bridge rectifier converts a sine wave into a positive or negative-going full-wave output. The polarity of the waveform will depend on the arrangements of the rectifier diodes. The circuit in *Figure 2.5a* will produce a positive-going output waveform.

As shown in *Figure 2.5a*, the bridge rectifier consists of four rectifier diodes. During the positive portion of the V_{sec} cycle, CR1 and CR3 are forward-biased, as shown in the equivalent circuit in *Figure 2.5b*. The current, i_1, will flow from terminal A to terminal B, producing a positive waveform; or one can say that the positive portion of the V_{sec} cycle is duplicated across

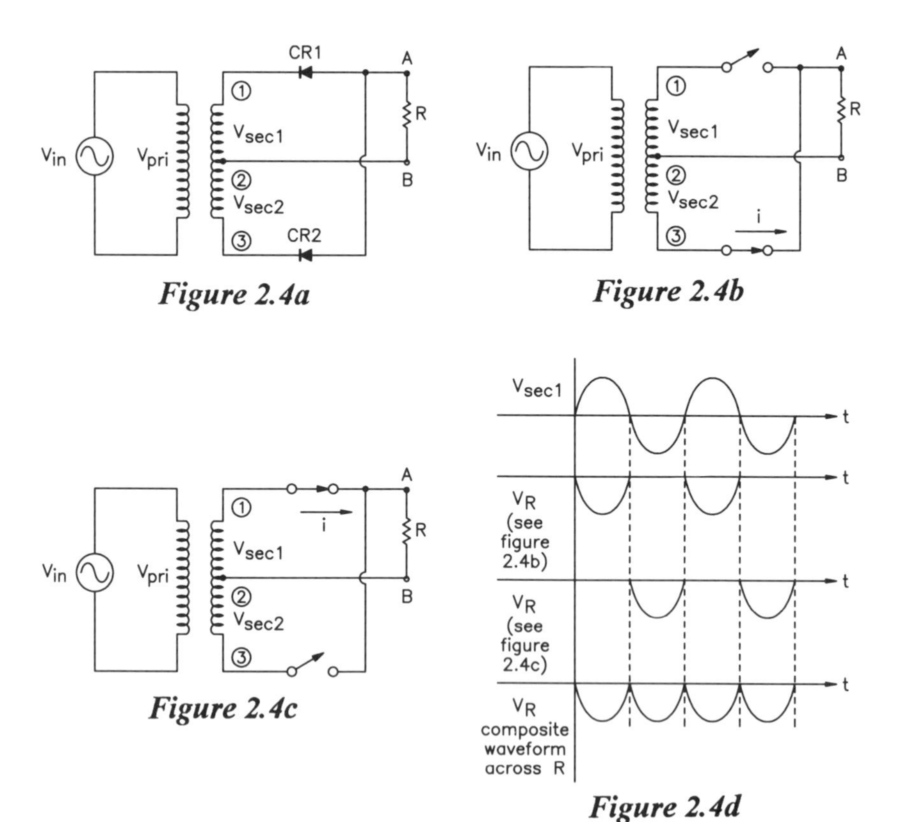

Figure 2.4a

Figure 2.4b

Figure 2.4c

Figure 2.4d

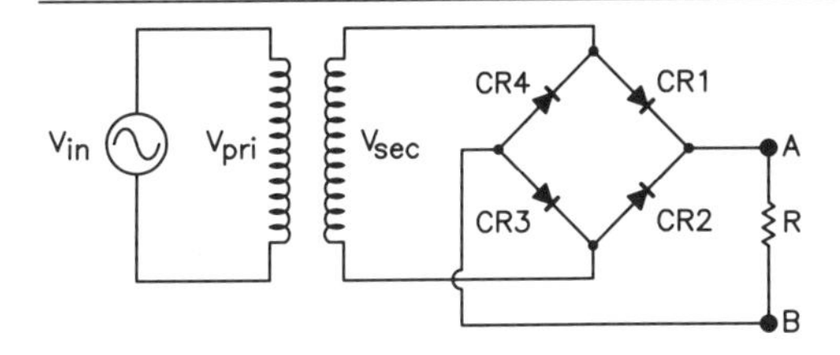

Figure 2.5a

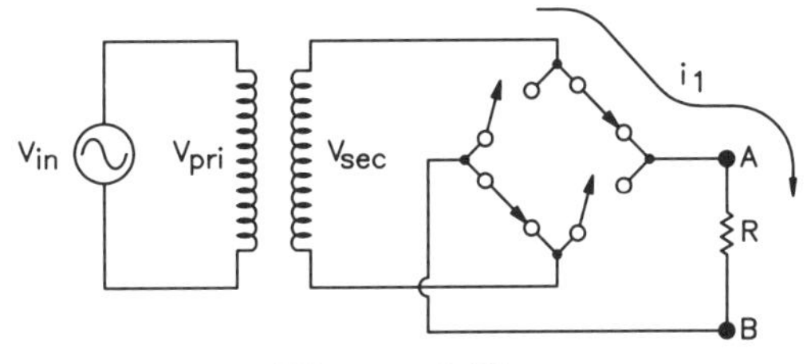

Figure 2.5b

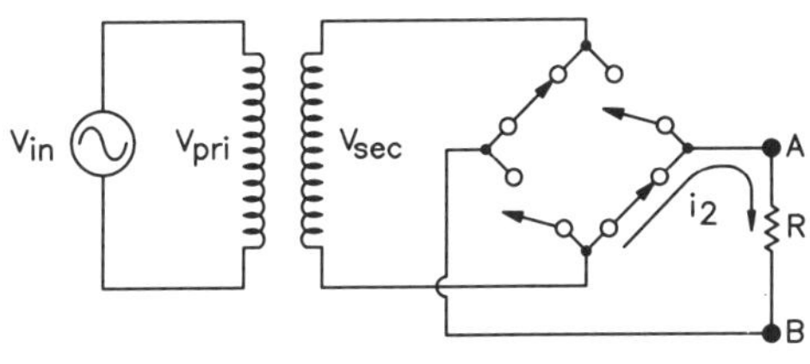

Figure 2.5c

R. During the negative portion of the V_{sec} cycle, CR2 and CR4 are forward-biased, as shown in the equivalent circuit in *Figure 2.5c*. The current, i_2, will flow from terminal A to terminal B, producing a positive-going pulse; or one can say that the negative portion of the V_{sec} cycle is "flipped" across R. (See *Figure 2.5d.*)

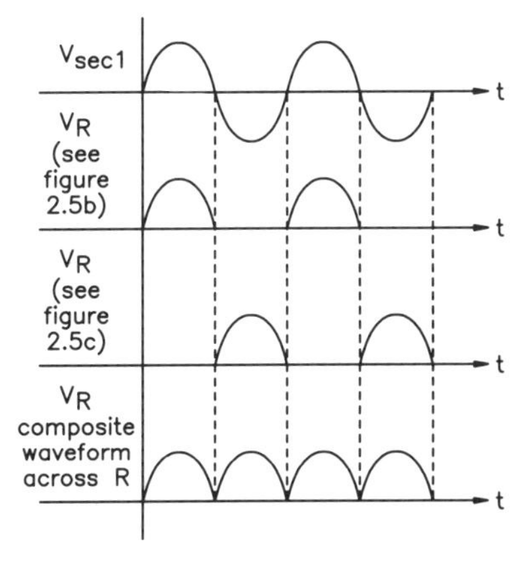

Figure 2.5d

If the diodes are configured as shown in *Figure 2.6a*, then the output will be a negative-going waveform. The reader may follow the above reasoning in deriving the negative-going output waveform. (See also *Figures 2.6b*, *2.6c*, and *2.6d*.)

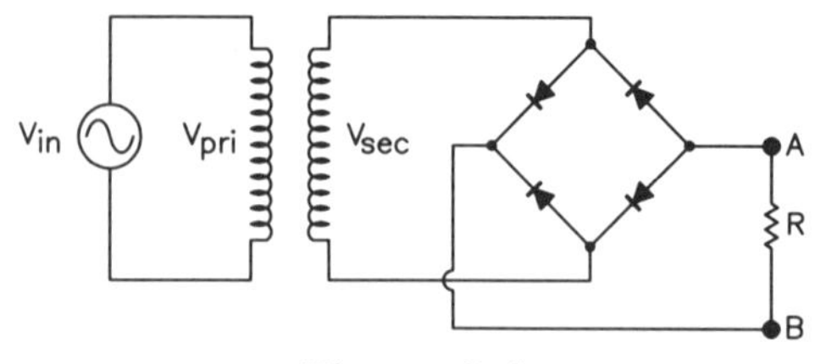

Figure 2.6a

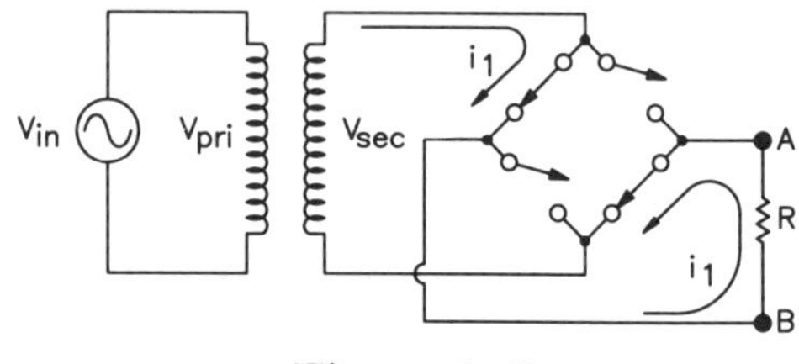

Figure 2.6b

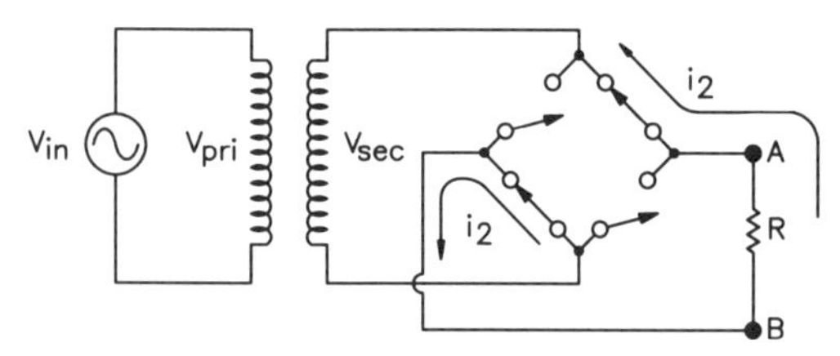

Figure 2.6c

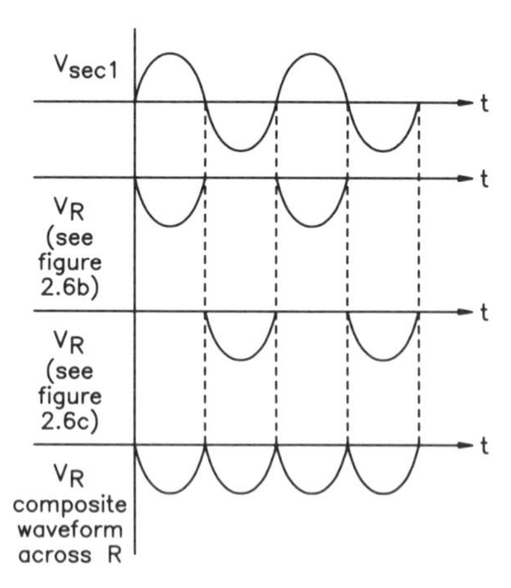

Figure 2.6d

2.2.2 FILTER

There are three types of commonly used filter circuitry:

Capacitor filter.
Resistor-Capacitor (RC) filter.
Inductor-Capacitor (LC) filter.

Capacitor Filter

Capacitor filters, shown in *Figure 2.7a*, are the most common filters used in present day switching power supplies. The main purpose of this filter is to smooth out rectified voltage. The rectified voltage is applied across the capacitor, which charges up to the peak voltage during the upward portion of the input voltage waveform, and during the downward portion of the wave, it supplies energy to the load. The capacitor continues to discharge until the next cycle, when the voltage level of the input voltage wave exceeds the discharged level of the capacitor, so the voltage across R_L appears as in *Figure 2.7b.*

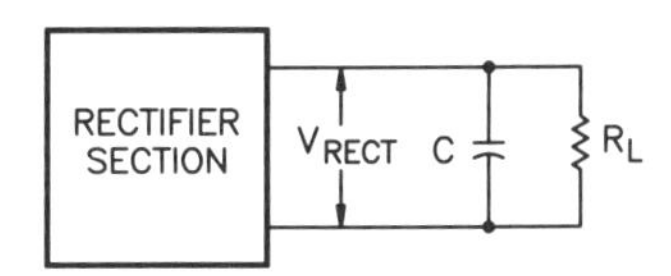

Figure 2.7a

The voltage waveform across R_L in *Figure 2.7b* consists of DC and AC portions. The AC portion is called the *ripple* voltage. The amplitude of the ripple depends on the values of the capacitor (C), the input frequency, and the load resistor (R_L). The ratio of the ripple voltage to DC voltage is called *ripple factor.* One disadvantage of using the capacitor is that while it is being charged, it draws high pulses of current.

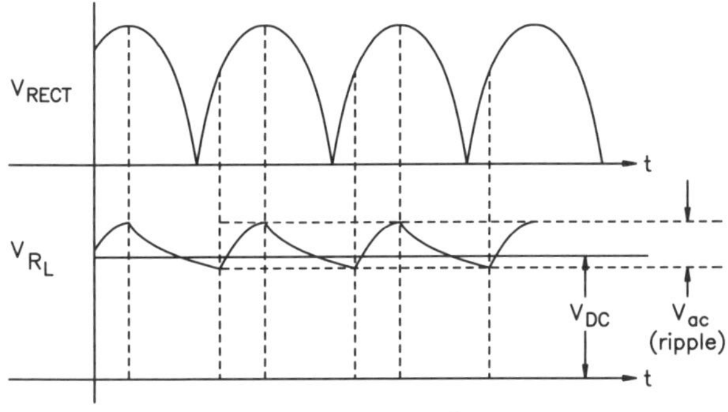

Figure 2.7b

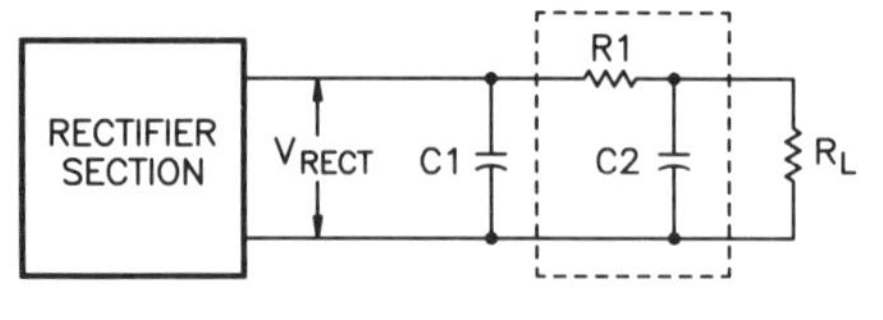

Figure 2.8

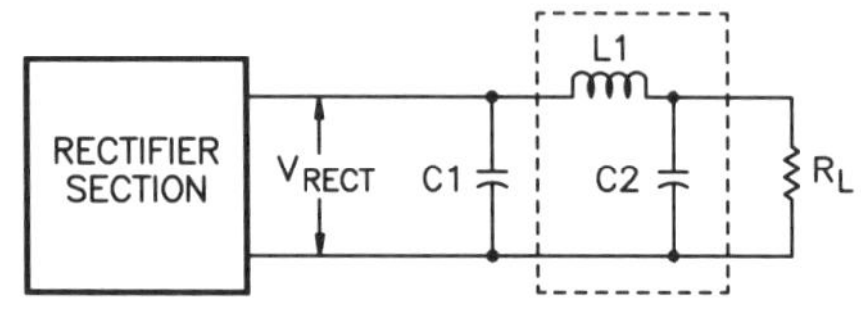

Figure 2.9

Resistor-Capacitor (RC) Filter

The ripple voltage can be reduced by adding another filter section after the capacitor filter mentioned in the previous section. This filter uses a resistor and an additional capacitor (see *Figure 2.8*).

When a proper combination of R1 and C2 is used, the RC filter will attenuate the ripple to a desired value. This filter is mainly used in the low power or low current applications. The reason for the low current application is that the current that passes through the resistor causes power (i * R1) dissipation, which reduces the overall efficiency of the power supply. The resistor also decreases the DC voltage because of the i * R1 drop across it. The ripple factor (AC ripple voltage/DC voltage) is lower than the capacitor filter (of the previous section). It is desirable to make R1 as small as possible to reduce i * R1 drop; but when R1 is reduced, the AC voltage (ripple) across the capacitor (and R_L) will be higher, and this will defeat the purpose of having an additional filter. In a high current application, i * R1 becomes significant, and the designer must concentrate on reducing R1 while making sure that the AC voltage across R_L does not increase.

Inductor-Capacitor (LC) Filter

In order to overcome some of the limitations of the RC filter, an LC filter can be used (see *Figure 2.9*). In the simplest way, one can say that the resistor (R1) in the RC filter can be replaced by a proper value of an inductor (L1), because L1 provides a low resistance while providing a high AC impedance; therefore, more of the DC voltage and less of the AC voltage will pass, creating a more desirable output of the filter section. The LC filter can also be called a two-pole or double-pole filter, while the RC filter is called a single-pole filter. The double-pole filter has a higher reduction or attenuation of the AC voltage than the single-pole filter. In general, it can be said that the higher the number of poles, the higher the attenuation; but there are other factors that become more critical for a higher order filter (or higher number of poles). Such discussion is beyond the scope of this book.

2.3 POWER FACTOR CORRECTION CIRCUITRY

The main purpose of the *power factor correction circuit* (PFC) is to correct the input current waveform so that it is in phase with the input voltage waveform. When the input voltage and current waveforms are in phase, the load seen by the input source is said to be purely resistive. Optimum power can be transferred to the load when it is resistive and not reactive, because reactive components cause the input voltage and current waveforms to lead or lag with respect to each other, which is undesirable. The reactive components are inductors and capacitors.

Power Factor

The power factor is defined as the ratio of real power to apparent power. The real power is the power that is dissipated by the system. For example, in the lab, this power would be measured on the input watt meter. The apparent power is the power calculated by multiplying the *rms* value of the input voltage and the *rms* value of the input current. The power factor of 1 is always seen in the purely resistive system (network). In the purely reactive network, the power factor is 0. Since most systems consist of resistive and reactive components, the power factor falls between the values of 0 and 1. For maximum power transfer, it is desirable to have a power factor of 1; therefore, most of the efforts in designing the PFC must go toward making the input load appear as a purely resistive load.

Power Factor Correction Circuit

The way to make the input load appear as a purely resistive load is to cancel out the reactive part of the system. For example, in the off-line switcher, the input section always contains a filter capacitor, making the input appear as a combination of resistive and capacitive load. An inductor of the same reactive value can be inserted in the circuit to cancel out capacitive reactance. This is called *passive power factor correction* technique. Though this is a valid technique, it has practical limitations. In the case of a high power system where a large filter capacitor may be used, it will require a large value of the input inductor, which may be physically large and heavy. Since in most power supply systems, it is desirable to have smaller components, this technique becomes unattractive. To overcome this problem, the technique of *active power factor correction* circuit is used.

Active Power Factor Correction (APFC) Circuit

The APFC circuit normally consists of a few components: a control IC chip; at least one transistor; at least one power diode; a high frequency inductor (much smaller and lighter than the one mentioned in the previous section); at least one filter capacitor; and a few discrete components, such as resistors, capacitors, signal diodes, etc. These components are arranged in a particular order called a topology. There are several topologies used for APFC. The two most common are the *Boost* and *Buck-Boost.* We will discuss these two topologies in greater detail in Chapter 4. The basic idea of the APFC is to make the load appear purely resistive to the input source, to align the input voltage and current waveforms with respect to each other.

CHAPTER 3
REGULATORS

3.1 INTRODUCTION

A regulator is a circuit which maintains a constant output. The output could be one of many different types; i.e. voltage, current, power, or some other parameter. For the purpose of explaining the basic concepts of regulators, this section will focus on *voltage* regulators only. (Readers may refer to other books for other types of regulators.) A voltage regulator converts a DC *input* voltage to a DC *output* voltage. The DC input voltage can be either a positive polarity or a negative one; however, we will focus on the positive input voltage, which could be either regulated or unregulated.

The block diagram of the basic voltage regulator is shown in *Figure 3.1a*, with *Figure 3.1b* showing a further breakdown. The basic idea is to maintain a desired value of V_O. V_O can be either a fixed value (i.e. +5V, -5V, +12V, +24V, etc.), or a variable value (i.e. 0—+5V, 0—+12V, etc.). The voltage V_{in} across terminals A and B is the input voltage to the G-block, and the voltage V_O across the terminals C and D is the output voltage of the G-block. The purpose of the G-block is to produce a desired value of V_O and deliver power to the load. The output load is connected across the V_O terminals, as shown in *Figure 3.1b*, and the purpose of the H-block is to monitor and see if the desired value of V_O is maintained. If V_O is higher or lower than the desired value, then the H-block sends a control signal to the G-block to adjust V_O. V_O can be either a fixed value or a variable value. In the case of a fixed V_O, no adjustment is needed, but for a variable V_O, it is necessary. This adjustment can be either manual or automatic (computer driven); however, in both cases this adjustment is performed in the H-block. The regulation is maintained under varying conditions of the input line and/or the output load.

V_O can be higher, equal, or lower than V_{in}. It can also be of opposite polarity, and the type of regulator circuit will determine the characteristics of it. More specifically, the circuit of the G-block will determine its characteristics. The techniques used in converting V_{in} to V_O can be categorized either as *linear* or *switching.*

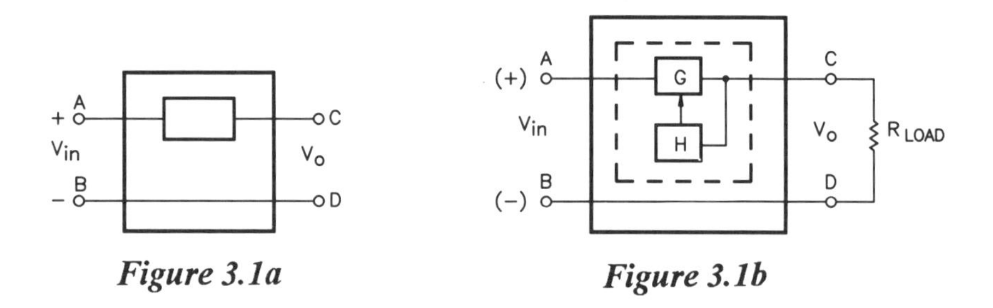

Figure 3.1a ***Figure 3.1b***

3.2 LINEAR REGULATORS

In the basic concept of a linear regulator, the output voltage, V_O, is kept at a desirable value by changing the voltage drop across the variable resistor (R_{series}) that is in series with the load (R_{load}). For example, in the circuit of *Figure 3.2a*, there is a variable resistor that is connected to the H-block. Let's assume that the resistance is changed by some signals from the H-block. Therefore, if the output voltage is less than the nominal output value, the H-block will adjust the resistance to create a lesser voltage drop across the variable resistor because $V_{Rseries} = I * R_{series}$, and increase the output voltage, V_O. If the output voltage is higher than the desired value, then the H-block will adjust the resistance to a higher value, thus creating a higher voltage drop across the variable resistor. In the real circuit, however, this variable resistor is replaced by a transistor, as shown in *Figure 3.2b*. The voltage drop across the collector to the emitter (V_{CE}) is controlled by the signal from the H-block and the signal from the H-block is the base current, I_B. I_B will control the collector current, I_C (recall that $I_C = \beta * I_B$, where β is the gain of the transistor), controlling the V_{CE} of the transistor, Q.

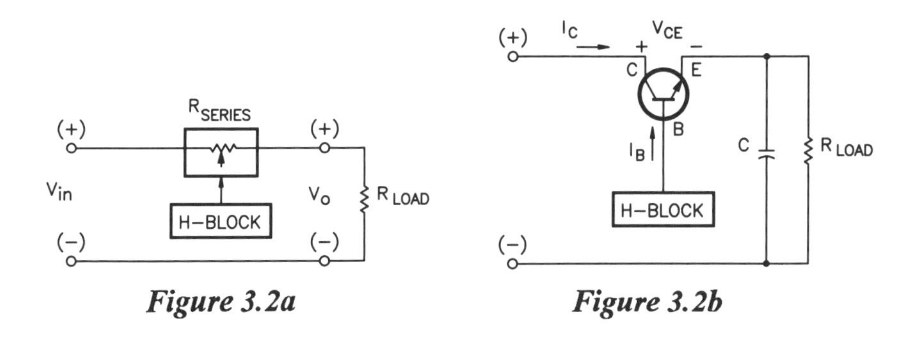

Figure 3.2a *Figure 3.2b*

3.3 SWITCHING REGULATORS

The technique in which the regulator maintains the output is called *switching*. In this technique, there are three commonly used devices; a diode, a transistor, and an inductor. On the output there is always a capacitor, which is used as a storage and filtering device. The three devices can be arranged in different combinations, but the most common arrangements are shown in *Figures 3.3a, b,* and *c*.

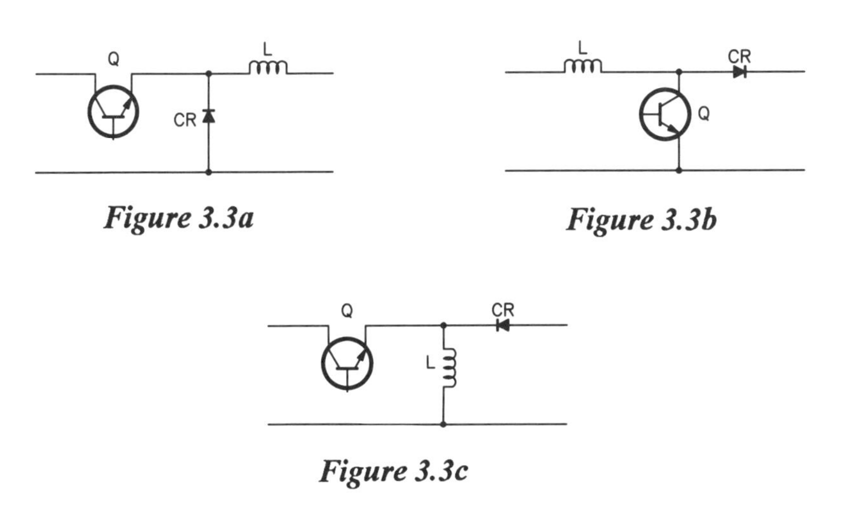

Figure 3.3a *Figure 3.3b*

Figure 3.3c

In the linear regulator, the output voltage is always lower than the input voltage; but in the switching regulator, the output voltage can be higher, equal, or lower than the input voltage. There are four commonly used switching regulators: Buck, Boost, Buck-Boost, and C'uk. The output voltage of the Buck regulator is always lower than the input voltage and the output voltage of the Boost regulator is always higher than the input voltage. The output voltage of the Buck-Boost and C'uk is either higher, equal, or lower than the input. (The polarities of both are opposite of the input.)

Switching

The switching device is either turned ON or OFF, operating in either an ON or OFF state. At this point, however, it is appropriate to mention that the consecutive ON and OFF states make up a switching cycle of period T, and we will call this a square wave or *pulse-width modulated* (PWM) switching waveform. The PWM will be described in detail in Chapter 5, where we will deal with a fixed frequency PWM waveform, meaning that the period of the cycle will remain the same, but the duration of ON time will vary. We will closely examine the circuit behavior during both states of a switching cycle. The transition state—changing from ON to OFF and from OFF to ON—also exists, but this state will not be discussed in this chapter because it is unnecessary in understanding the basic concepts of the switching regulators. The following subsections explain the four switching regulator topologies in detail.

3.3.1 BUCK REGULATOR

According to Webster's Dictionary, the word "buck" means "to break into small bits." In the circuit configuration of *Figure 3.4*, the output voltage is a "smaller bit" of the input voltage; therefore, it is called a "Buck" configuration. This bucked output voltage is regulated by the circuit; therefore, it is called a *Buck regulator*. This regulator also produces output voltage of the same polarity as the input voltage. For example, for a positive input voltage, it produces a positive output voltage, and for a negative input voltage, it produces a negative output voltage. The G-block of *Figure 3.1b* will contain the circuit of *Figure 3.4*. The V_O of the G-block will be the bucked output voltage. Let's look at the circuit of *Figure 3.4* in detail to see how this bucked output voltage is produced. We will examine the circuit operation during the ON and OFF states.

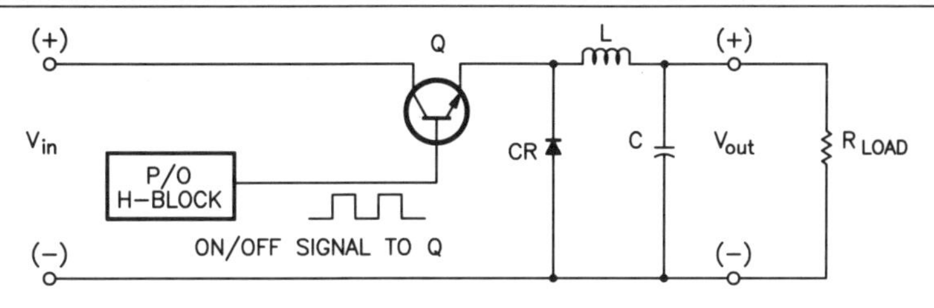

Figure 3.4

ON State

In *Figure 3.4*, when Q is turned ON, it acts as a short and causes V_{in} to appear at point A. We will call this voltage V_A. Since V_A is positive, and the cathode of CR is connected to this point, the diode is reverse-biased; therefore, it acts as an open and is not considered a part of the circuit in this state. (The equivalent circuit for this is shown in *Figure 3.5a.)* Between the input and the output, only the inductor appears in series. Let us see the effect of the location of the inductor on the output voltage.

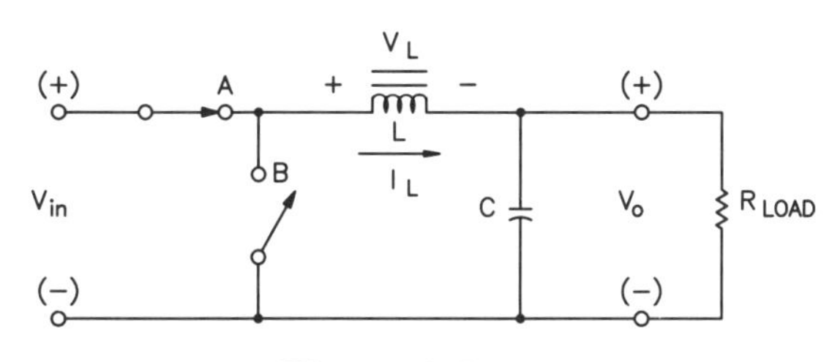

Figure 3.5a

It is known from basic electronic theory that, initially, when the voltage is applied across an inductor, it will act as an open. The current in the circuit will charge the inductor, and as the inductor is charging, the voltage across it will decay. In the equivalent circuit, the voltage applied across the inductor is the difference of V_O and V_A, or $V_L = V_A - V_O$, where V_L is the voltage across the inductor. (We will assume that the circuit is in a steady state, or it has been operating normally for some time and some level of V_O exists.) When Q is initially turned ON, V_L will appear across the inductor, the current in the series circuit will start flowing, and the current will begin to charge the inductor. Let us manipulate the above equation differently to see how V_L affects V_O by rewriting the equation as follows: $V_O = V_A - V_L$. It can be seen that because of the voltage drop across the inductor, V_O will always be lower than V_A. We can also see that by changing the voltage across the inductor, we can change the value of V_O. In the Buck regulator, V_O is regulated by controlling the voltage drop across the inductor.

Now let us see how we can change the value of V_L. We know that if we continue to charge the inductor, the voltage will decay and V_L will become smaller; therefore, V_O will be higher, but V_O can never be higher than V_A.

There are several ways to control the voltage across the inductor. Many designers refer to the V_L drop as $L * d_i/d_t$. L is the inductor value, d_i is a mathematical way of saying the difference between two levels of inductor current (for example, $i_1 - i_2$ in *Figure 3.6*), and d_t is the difference in two different points in time (for example $t_2 - t_1$ in *Figure 3.6*). The correct mathematical way to represent the inductor voltage is $V_L = L * d_i/d_t$. When d_i is divided by d_t, we get the slope of the current waveform, as shown in *Figure 3.6*; therefore, the voltage across the inductor is L multiplied by the slope of the inductor current. We can change V_L by controlling any of the three variables; however, during the live operation of the circuit, it is much easier to change the slope of the inductor current than the value of L.

The slope is controlled by changing the time difference in *Figure 3.6*. This time difference is called the ON time (or t_{on}) of Q, and we can see that t_{on} controls V_L, and that in turn controls V_O. Therefore, by controlling the ON time of the Q, we can control the output voltage, V_O, for the ON state of the circuit operation.

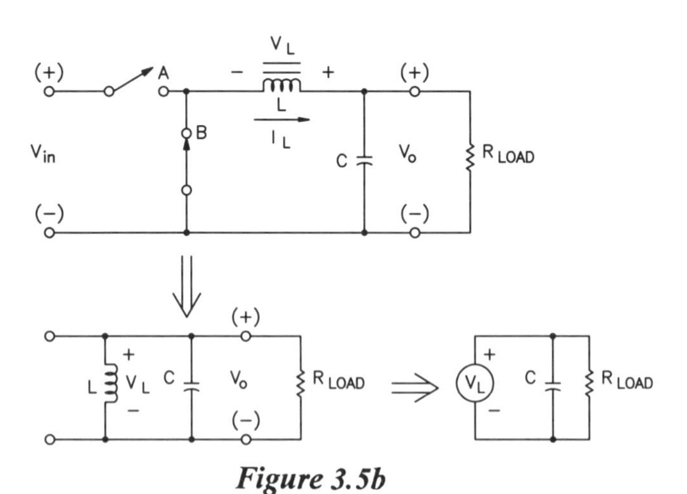

Figure 3.5b

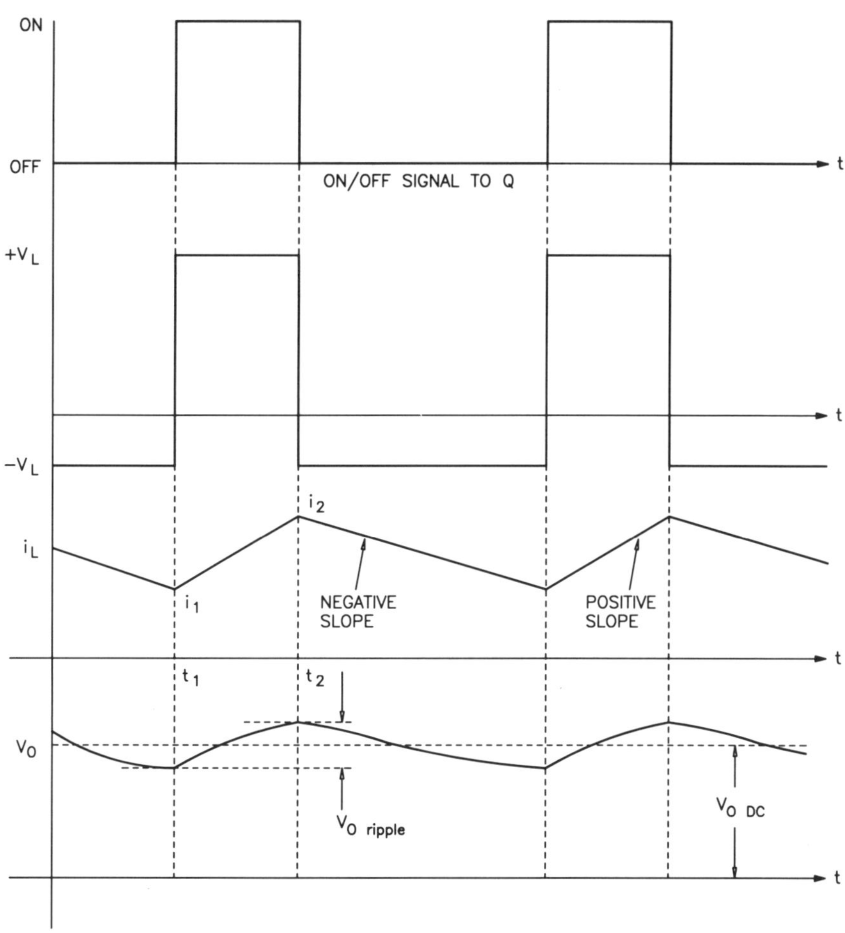

Figure 3.6

OFF State

When Q is turned OFF, the input voltage, V_{in}, is disconnected from the circuit, and the voltage, V_L, across the inductor reverses the polarity; therefore, the voltage at point A will become negative with respect to V_O. When V_A becomes negative, the diode, CR, will become forward-biased, and point A will be at ground potential. (The equivalent circuit for this is shown in *Figure 3.5b.*) The voltage of the inductor appears across the output section, and the stored energy is delivered to the output load; therefore, the inductor can be shown as an energy source to the output load in the OFF state. Though the voltage across the inductor is reversed, the current moves through it in the same direction. However, the slope of the current will be negative because the inductor is now discharging the current into the load, as opposed to charging it during the ON time, and the discharging of the inductor will continue until Q is turned back ON.

At this point, we can say that one switching cycle is complete. The charging (during the ON state) and the discharging (during the OFF state) action creates the voltage ripple at the output capacitor C. As shown in *Figure 3.6*, the final output voltage of V_O is the average of these voltages. Mathematically, this voltage can be represented with the following equation: $V_O = (t_{on}/T) * V_A$, where T is the period of one switching cycle (or we can say the duty cycle $D = t_{on}/T$ and $V_{in} = V_A$); therefore, we can rewrite the output voltage equation: $V_O = D * V_{in}$. In the lab, this voltage will be measured as DC volts on the voltmeter.

Advantages

High efficiency.
Simple.
Easy to stabilize.
High frequency operation.
Easy to stabilize the loop.

Disadvantages

No isolation between input and output.
Protection circuit is needed in case the Q shorts.
The input current is discontinuous.

3.3.2 BOOST REGULATOR

The word "boost" means "increase." In the circuit configuration of *Figure 3.7*, the output voltage is boosted from the input voltage; therefore, it is called a *Boost configuration.* This boosted output is regulated by the circuit; therefore, it is called a *Boost regulator.* This regulator also produces output voltage of the same polarity that is in the Buck regulator. The G-block of *Figure 3.1b* will contain the circuit of *Figure 3.7*, and the V_O of the G-block will be the boosted output voltage. Let's look at the circuit of *Figure 3.7* in detail to see how this boosted output is produced. We will examine the circuit operation during the ON and OFF states.

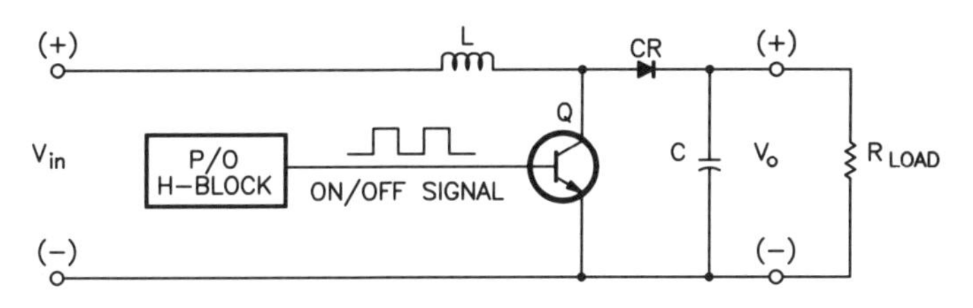

Figure 3.7

ON State

In *Figure 3.7*, when Q is turned ON, it acts as a short and causes point B to be at ground potential. Since the anode of the diode is connected to this point, the diode is reverse-biased; therefore, it acts as an open and disconnects the output load. (The equivalent circuit for this is shown in *Figure 3.8a.*) The entire V_{in} appears across the inductor, and the current from V_{in} will start flowing through the inductor, charging it. The slope of this current will be positive, and the direction will be from V_{in} to ground. While the inductor is charging, the output capacitor, C, is discharging into the load. (As in the Buck regulator, we will assume that the circuit is in a steady state and some value of V_O is present.) In this state, the capacitor will continue to supply the power to the load.

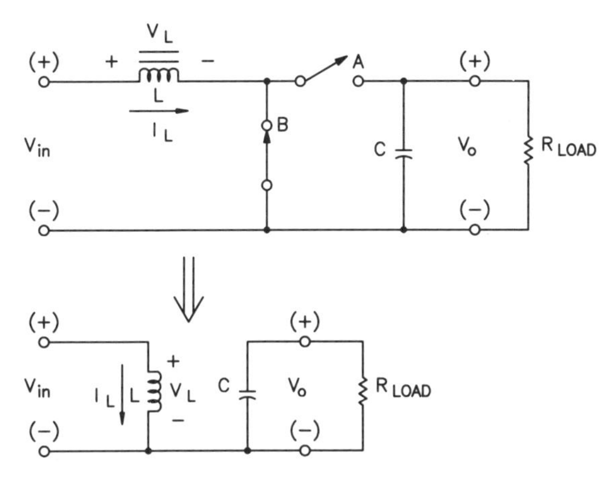

Figure 3.8a

OFF State

When Q is turned OFF, point B is disconnected from the ground potential and becomes positive. When the voltage at this point exceeds the capacitor voltage, V_O, then the diode will be forward-biased and start conducting. Also, the polarity of the inductor voltage will be reversed, but the current will continue to flow in the same direction, and it will have a negative slope because it is discharging its current into the output load. (The equivalent circuit of this is shown in *Figure 3.8b*.) The voltage across the inductor is in series with the input voltage; therefore, V_O is the sum of the two voltages, or $V_O = V_{in} + V_L$. V_O will always be higher than V_{in}, because the polarity of V_L is always the same as V_{in}.

At this point, one switching cycle is complete. The charging and the discharging action creates the voltage ripple at the output capacitor, C. As shown in *Figure 3.9*, the final output voltage, V_O, is the average of these voltages. Mathematically, voltage V_O can be represented with the following equation:

$$V_O = \left(\frac{1}{1-\left(\frac{ton}{T}\right)}\right) * V_A$$

, where T is the period of one switching cycle.

Or we can say the duty cycle $D = t_{on}/T$ and $V_{in} = V_A$; therefore, we can rewrite the output voltage equation to:

$$V_O = \left(\frac{1}{1-D}\right) * V_{in}$$

In the lab, this voltage can be measured as DC volts on the voltmeter.

Advantages

Provides a step up voltage without a transformer.
High efficiency.
Simple.
High frequency operation.
The input current is continuous.

Disadvantages

No isolation between input and output.
High peak current in the switching device, Q.
Poor transient response.
Only one output is possible.
Difficult to stabilize the loop.
Protection circuit required for the switching device.

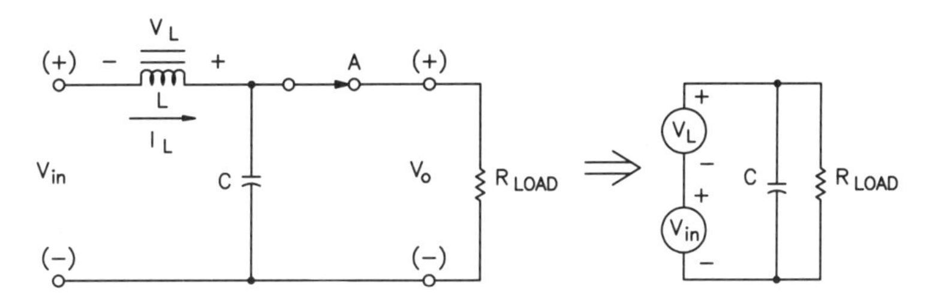

Figure 3.8b

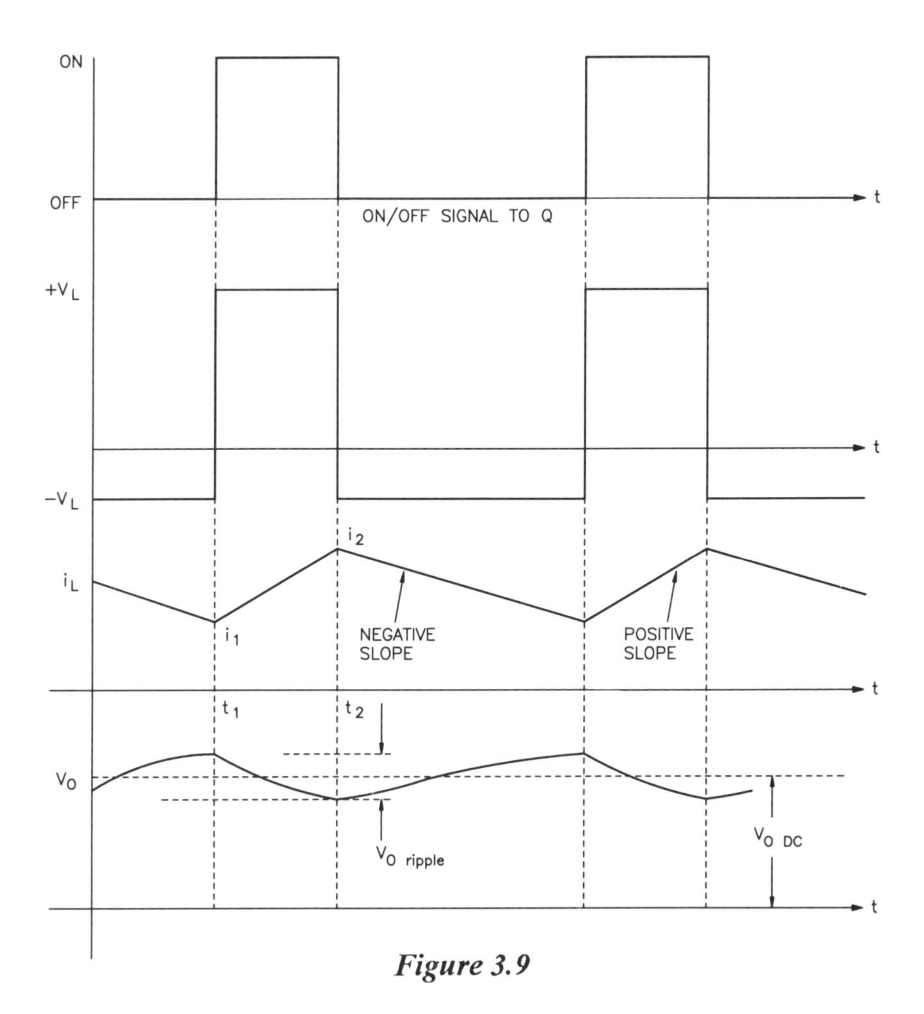

Figure 3.9

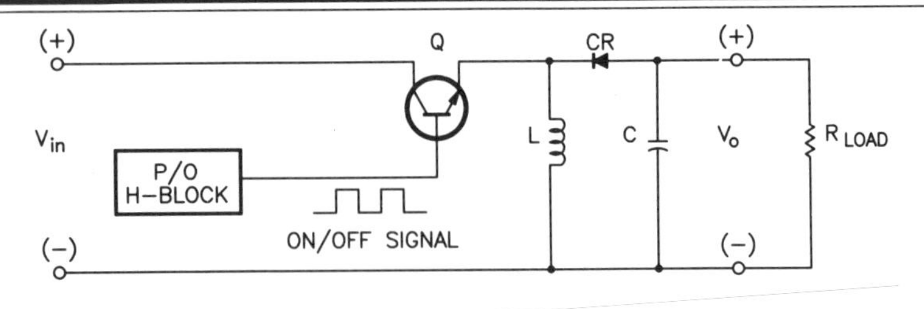

Figure 3.10

3.3.3 BUCK-BOOST REGULATOR

This regulator combines the concepts of the Buck and the Boost regulators, and its output voltage can be either higher, equal, or lower than the input voltage. One significant difference from the previous two concepts is that the polarity of the output voltage is always the opposite of the input voltage. For example, for the positive input voltage, a negative output voltage is produced, and for the negative input voltage, a positive output voltage is produced. The G-block of *Figure 3.1b* will contain the circuit of *Figure 3.10*. The V_O of the G-block will be either higher, equal, or lower than V_{in}. Let's look at the circuit of *Figure 3.10* in detail to see how the output voltage is produced. We will examine the circuit operation during the ON and OFF states just as we did with the Buck and Boost regulators.

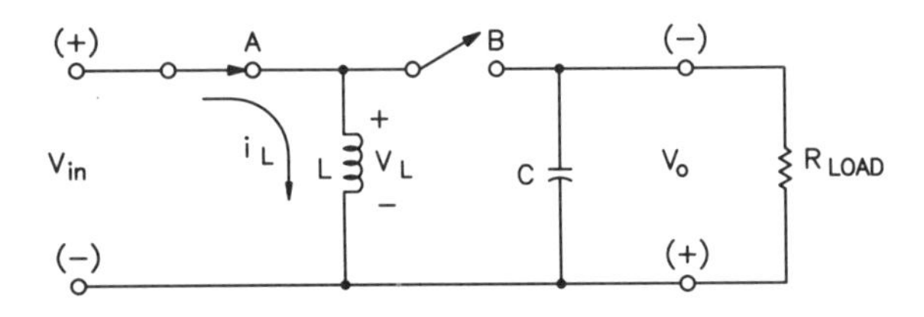

Figure 3.11a

ON State

In *Figure 3.10*, when Q is turned ON, it acts as a short and causes the entire V_{in} to appear at point A. Since V_{in} at point A is positive, and the cathode of CR is connected to this point, the diode is reverse-biased. The diode will then act as an open and disconnect the output circuit from the input. (The equivalent circuit of this state is shown in *Figure 3.11a*.) The equivalent circuit is the same as the Boost configuration (see *Figure 3.8a*); therefore, the circuit operation is the same as the Boost regulator in the ON state. The current from V_{in} will flow through the inductor and charge it. The direction of this current is from V_{in} to ground; therefore, the polarity across the inductor appears as shown in *Figure 3.11a*, and the slope of the current will be positive. While V_{in} is charging the inductor, the output capacitor is discharging into the load. (As with the previous two regulators, we will assume that the circuit is in a steady state and some value of the negative V_O is present.) The capacitor will continue to supply the power to the load.

OFF State

When the Q is turned OFF, the input voltage, V_{in}, is disconnected from the circuit. This is the same operation that we observed in the Buck regulator. The equivalent circuit for this state is shown in *Figure 3.11b*, and it can be seen that the polarity of the voltage across the inductor is reversed; therefore, the voltage at point B will become more negative with respect to V_O. When the voltage at point B becomes more negative than V_O, the diode will become forward-biased and act as a short, connecting point B to the output circuit. The inductor now appears across the output circuit. The energy stored in the inductor during the ON state is now transferred to the load.

Now let us see how the output voltage is higher, equal, or lower than the input voltage (see *Figure 3.12*). Recall that in the Buck regulator, the output voltage, V_O, was achieved by the following equation: $V_O = D * V_{in}$, where D is the duty cycle, and $D = t_{on}$/period of the switching cycle. Also, we saw that the output of the Boost regulator was achieved by $V_O = (1 / (1 - D)) * V_{in}$. In the Buck-Boost configuration, we can combine the equations to achieve the following:

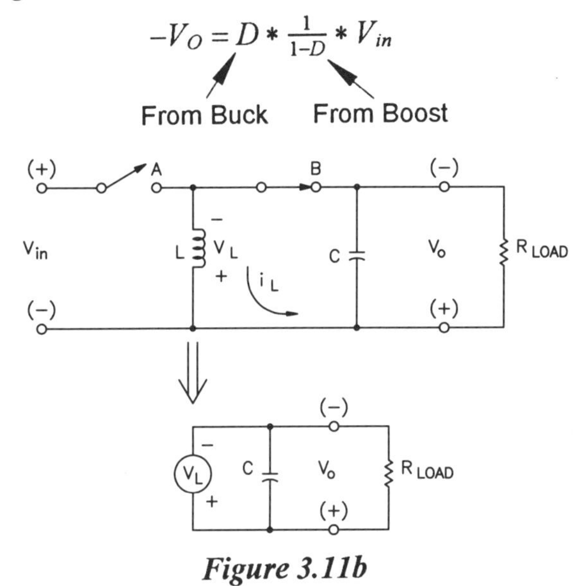

Figure 3.11b

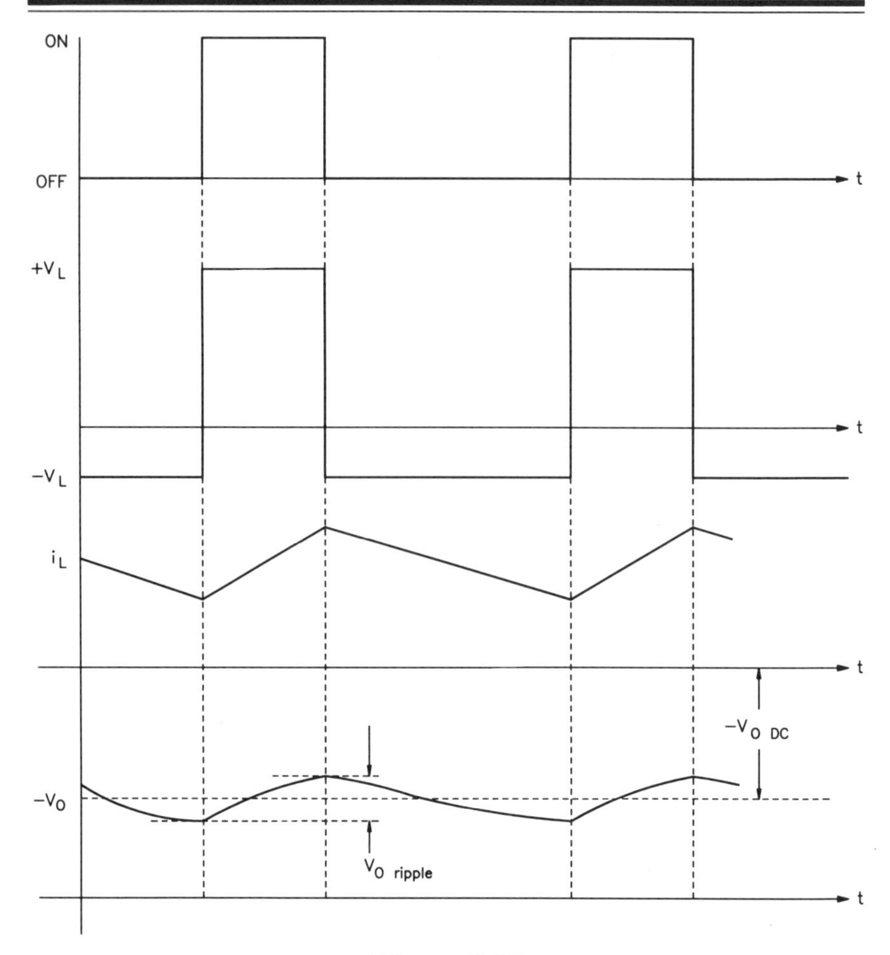

Figure 3.12

From the above equation, the output voltage can be controlled by changing the duty cycle, D, and depending upon the value of D, the output voltage can be higher, equal, or lower than the input, V_{in}.

Advantages

Simple.
High frequency operation.
Reverses the output voltage without a transformer.
High efficiency.
Easy to implement current sensing circuit to protect from overcurrent and short circuit condition.

Disadvantages

No isolation between input and output.
The input current is not continuous.
The switching device carries high peak current.
Poor transient response.
Only one output is possible.

3.3.4 C'UK REGULATOR

Unlike the previous three regulator topologies, this regulator derives its name from its inventor, Dr. Slobodan C'uk. It is also known as a "Boost-Buck regulator," and similar to the Buck-Boost regulator, the output voltage can be higher, equal, or lower than the input voltage. In addition, the output voltage is opposite in polarity to the input voltage. In fact, the mathematical expression to calculate the output voltage, V_O, is the same equation as the Buck-Boost regulator. The G-block of *Figure 3.1b* will contain the circuit of *Figure 3.13*. The V_O of the G-block will be either higher, equal, or lower than the V_{in}. Let's look at the circuit of *Figure 3.13* in detail to see how the output voltage is produced. As with the previous three regulators, we will examine the circuit operation during both states; however, we will first look at the OFF state and then the ON state.

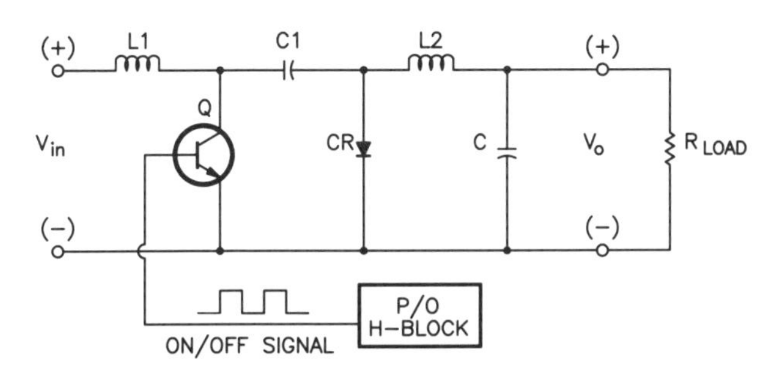

Figure 3.13

OFF State

In *Figure 3.13*, when Q is turned OFF, the current from the input voltage source, V_{in}, will flow through L1, C1, and the diode. The anode of the diode is positive; therefore, it can be replaced by a short, bringing point B to ground potential. The equivalent circuit of this state is shown in *Figure 3.14a*, and we can look at it in two sections. One section contains L1 and the other section L2. We will assume that the circuit is operating in the steady state and L1 and L2 are previously charged; therefore, when I_{L1} is flowing, both the input source V_{in} and L1 is supplying energy to C1. As C1 is charging up, I_{L1} is decreasing in the circuit; therefore, it will have a negative slope, as shown in *Figure 3.15*.

Now let's look at the other section of the equivalent circuit. In this section, the energy stored in L2 will supply power to the load and in the process, discharge; therefore, current I_{L2} will decrease. While it is being discharged, the voltage across it and V_O will decay. (This decayed voltage creates part of the ripple. The other part is the rising of this voltage during the ON time.) In this state both I_{L1} and I_{L2} will have a negative slope.

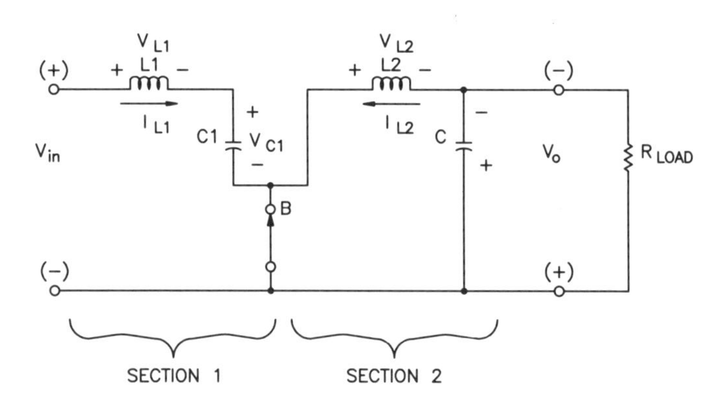

Figure 3.14a

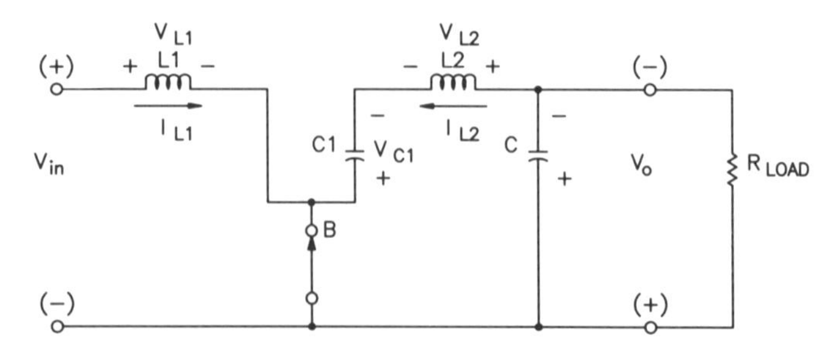

Figure 3.14b

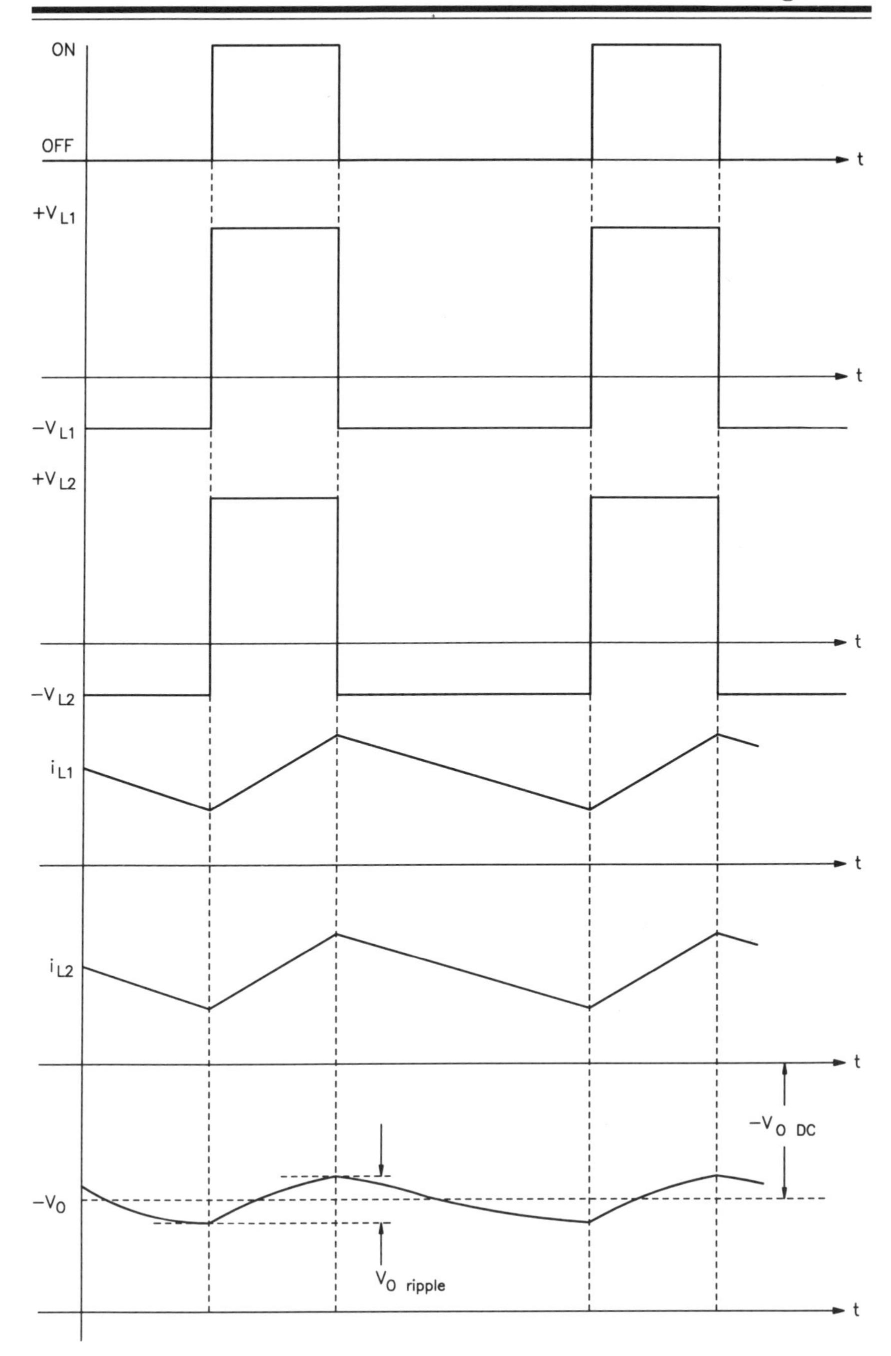

Figure 3.15

ON State

When Q is turned ON, the entire V_{in} will appear across L1. The current will flow from V_{in} to L1 and charge it, and the energy stored in L1 during this process will be used to charge up C1 in the next OFF state. In the other section, the anode of the diode is connected to the negative terminal of C2; therefore, it will be reverse-biased and act as an open. Since V_O and V_{L2} are less than V_{C1}, the current will start flowing from C1 through L2 to the load, and back to C1 through the short (forward-biased diode). This current will charge up L2, and stored energy in L2 will be delivered to the output load during the next OFF state.

As mentioned earlier in the section, the output voltage, V_O, can be achieved by the same mathematical expression as the Buck-Boost regulator:

$$-V_O = D * \frac{1}{1-D} * V_{in}$$

From Buck   From Boost

As described in the previous section, the duty cycle, D, will control the final output voltage, and the relative magnitude of V_O will depend on its value.

Advantages

This converter has the highest efficiency.
Input and output currents are continuous.
Low switching losses.

Disadvantages

The switching device sees a high peak current.
The output capacitor sees a high ripple current.

CHAPTER 4
CONVERTERS

4.1 INTRODUCTION

In the previous chapter, we looked at various types of regulators. In each of those circuits, the input and output are measured with respect to a common terminal; therefore, no isolation exists between the input (source) and the output (load). Also, we talked about certain characteristics of the input and output. For example, in the Buck regulator, the input must always be higher than (and of the same polarity as) the output, and in the Boost regulator, the input must always be lower than (and of the same polarity as) the output. Also, only a single output is possible. To overcome these limitations, a new set of circuits is developed. These circuits simply add a transformer and place the switch (Q) in a different place. The transformer provides the isolation between the input and the output, and also provides the voltage scaling and removes the restrictions on polarity. In the following sections, we will examine these features in detail.

These new circuits are called *converters*. Although many texts and designers use the term "converter" for both the isolated and non-isolated circuit configurations, this book chooses to use the term "regulator" for the non-isolated circuit and "converter" for the isolated version. It must be made clear to the reader that both types of circuits convert and regulate.

In the process of developing the converter circuits, many different types of configurations have been derived, but most of them are variations of the following three circuits:

1) The forward converter (also known as the Buck converter or step-down converter).
2) The flyback converter (also known as the Buck-Boost converter or the step-down/step-up converter).
3) The push-pull converter (also known as the Buck derived converter).

In the basic concept of the forward and flyback converters, a single switching device (Q) is used, while in the push-pull converter, two or more switching devices are used (see *Figure 4.4a*). The push-pull converter can be arranged in several ways, but the most popular ones are the center tapped, the half-bridge, and the full-bridge.

The center tapped is often referred to as a push-pull. Throughout this book, we will use the term push-pull when referring to this topology. The push-pull topology has a center tapped winding on the primary side of the transformer. Both half-bridge (see *Figure 4.5a*) and full-bridge (see *Figure 4.6a*) topologies have a single primary winding with no center tap. The push-pull and the half-bridge topologies use two switches (Q1 and Q2), while the full-bridge uses four switches (Q1 ,Q2, Q3 and Q4). The switches in the converters are turned ON and OFF as in the regulator sections; therefore, we will examine the circuits operating in the ON state and the OFF state. These converter circuits can provide multiple outputs.

Transformer Basics

A transformer is made up of at least two windings; primary and secondary. The primary and secondary windings have a number of turns, N_P and N_S respectively. These windings are also known as coils, and are wound on a core. The direction of the windings determines the positions of the dots in *Figure 4.1a*. The coils can be divided into smaller sections. For example, a coil with 10 turns can be broken down into two 5 turn coils, as shown in *Figure 4.1b*. If we connect terminal *b* and terminal *c* together, the connection is called a center tap because we are tapping at the center of the coil. The input is normally applied to the primary side and output is on the secondary side. The voltage and current on the secondary side depends on the ratio of N_S to N_P, as shown below:

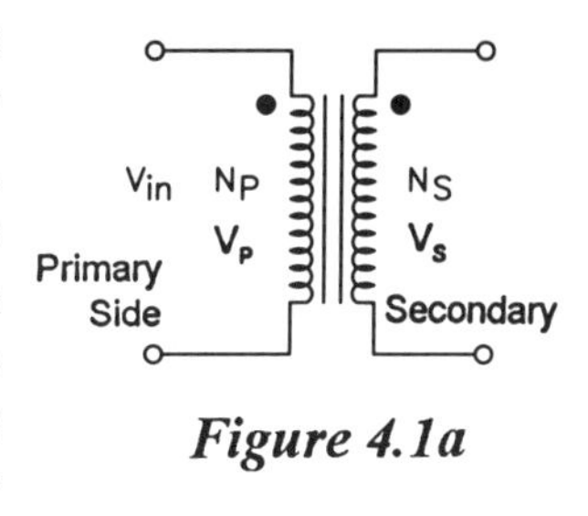

Figure 4.1a

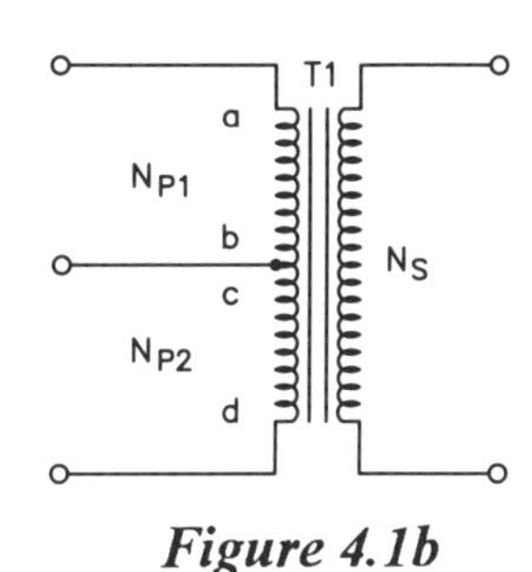

Figure 4.1b

$$\frac{Vs}{Vp} = \frac{Ns}{Np}$$

To find the secondary voltage (V_S), we can rearrange the above equation to:

$$Vs = \frac{Ns}{Np} * Vp$$

$$\frac{Is}{Ip} = \frac{Np}{Ns}$$

To find the secondary current (I_S), we can rearrange the above equation to:

$$Is = \frac{Np}{Ns} * Ip$$

The turns ratio in the current expression is $\frac{Np}{Ns}$ as opposed to $\frac{Ns}{Np}$ in the voltage expression.

For example, let us say that we want the secondary voltage of 30V peak when the input voltage is 10V peak. In the above equation, we can substitute 30 for V_S and 10 for V_P, then from this ratio we can get the turns ratio:

$$\frac{Ns}{Np} = \frac{Vs}{Vp} = \frac{30}{10} = 3$$

Therefore, the turns ratio is 3. In other words, the secondary coil has 3 times more turns than the primary coil; so if the primary has 15 turns, then the secondary will have 45 turns, etc. From the equation for the current, it can be seen that the secondary current (I_S) will be 3 times less than the primary current:

$$\frac{Is}{Ip} = \frac{Np}{Ns} = \frac{1}{3} = 0.333$$

So for the primary current of 60A, the secondary current will be 20A.

4.2 SWITCH-MODE CONVERTERS

4.2.1 FORWARD CONVERTER

The forward converter is an isolated version of the Buck regulator. An ideal circuit for this topology is shown in *Figure 4.2a.* Recall that in the Buck regulator (see *Figure 3.4* in Chapter 3), there are four main components; switch (Q), diode (CR), inductor (L), and capacitor (C). In the forward converter configuration, all four components remain, but two extra components, T1 and CR1, are added. (Notice that CR is now labeled as CR2 and L is labeled as L1.) T1 is providing an isolation and also the voltage scaling. The voltage scaling may be necessary, because as was mentioned in the Buck regulator section, the input voltage, V_{in}, must be higher than the output voltage, V_O; and when the input voltage is lower than V_O, the turns ratio (N_S to N_P) of the transformer can be changed to achieve a higher voltage on the secondary side. The polarity (or dot) positions in the forward converter are always as shown in *Figure 4.2a.* In the Buck regulator, the ON/OFF pulses appeared in the circuit because of the switching action of Q. In this case, the switching action of Q couples through T1 and the secondary of T1 acts as the source of these pulses in the circuit. We will now examine the circuit operation during the ON and OFF states.

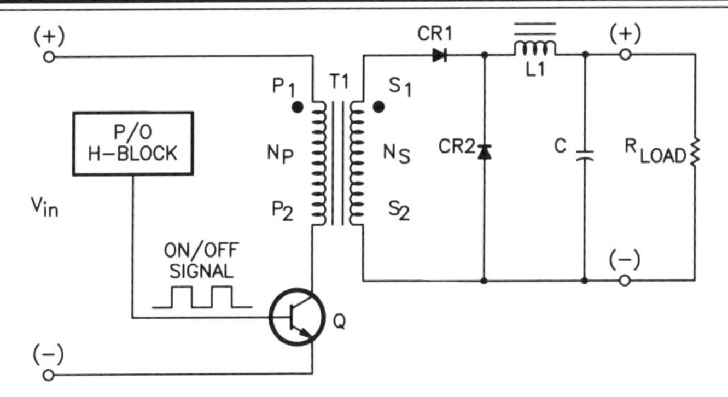

Figure 4.2a

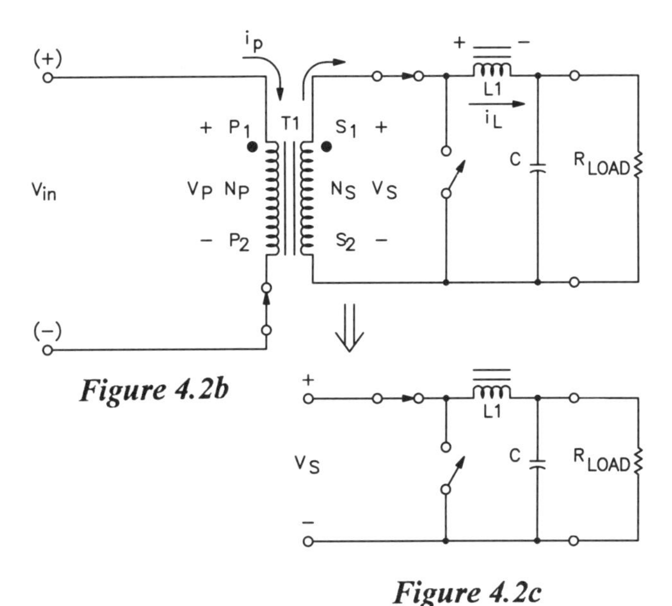

Figure 4.2b

Figure 4.2c

ON State

In *Figure 4.2a*, when Q is turned ON, it acts as a short and causes V_{in} to appear across N_p, as shown in the equivalent circuit in *Figure 4.2b*. The V_{in} across the primary turns (N_P) will cause i_p to flow through the primary side, charging up the primary coil. This energy is coupled to the secondary side. i_p enters the P1 (dot) point and leaves the P2 point, creating a positive voltage polarity at the P1 (dot). Recalling the rules of the *dot* conventions, the same polarity will appear on the dotted sides of the coils—in our case, it is the terminal S_1. When S_1 is positive, CR1 becomes forward-biased, acting as a short, and CR2 acts as an open because it is reversed-biased.

Let us redraw the circuit of *Figure 4.2b* as the circuit of *Figure 4.2c*. The circuit is now similar to the circuit of *Figure 3.5a*. In Chapter 3, the detailed operation of this circuit can be found in section 3.3.1

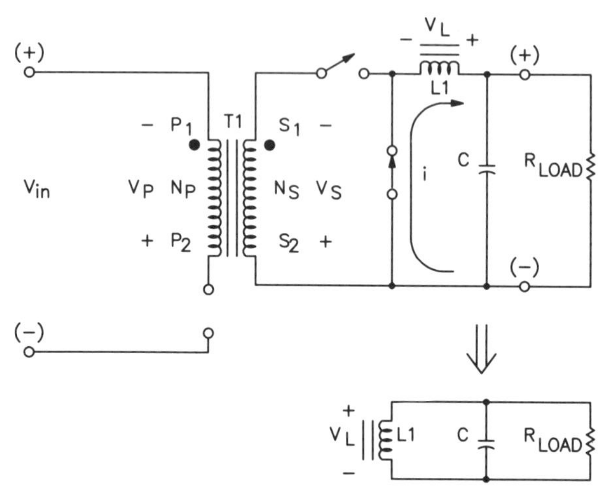

Figure 4.2d

OFF State

When Q in *Figure 4.2a* is turned OFF, creating an open in the primary side of the circuit, the current, i_p, stops flowing. Due to the characteristics of the transformer, the magnetic field across the primary will start collapsing and reverse the polarity. P_1 (dot) will now become negative and P_2 will be positive. On the secondary side, the S_1 (dot) terminal will become negative and S_2 will be positive. The negative voltage at the S_1 terminal will reverse-bias CR1; therefore, CR1 will act as an open. As in the primary side of the transformer, the magnetic field across L1 will start collapsing and the voltage across it will be reversed. This will forward-bias CR2. (The equivalent circuit is drawn in *Figure 4.2d.* The circuit in this figure is similar to the one in *Figure 3.5b*. Refer to section 3.3.1 for the detailed operation of this circuit.)

This completes one switching cycle; however, let's stop for a moment and see what happens in the OFF state, when the voltage polarity across the primary coil reverses. In the equivalent circuit of *Figure 4.2d*, it can be seen that when Q is in the OFF state, the voltage across Q (V_Q) is the sum of V_{in} and V_P, or $V_Q = V_{in} + Vp$ (see *Figure 4.1e*). This voltage creates a stress across Q that is undesirable because it may degrade the performance of Q, or cause it to fail. To reduce the stress, the stored energy (also called the trapped energy) in the inductor must be removed. To achieve this, a "clamp winding" and an additional diode are added to the original circuit of *Figure 4.2a*. The circuit of *Figure 4.2a* can be redrawn as the one in *Figure 4.2f.* Let's examine the ON and OFF states with these additional components. Notice the position of the dot in the clamp winding.

ON State with Clamp Winding

In *Figure 4.2g*, when P_1 (dot) is positive, S_4 (dot) is also positive, making CR3 reversed-biased; therefore, it acts as an open, and the clamp winding circuit has no effect in this mode of the circuit operation. (The equivalent circuit of *Figure 4.2g* is also similar to the one in *Figure 4.2b*.)

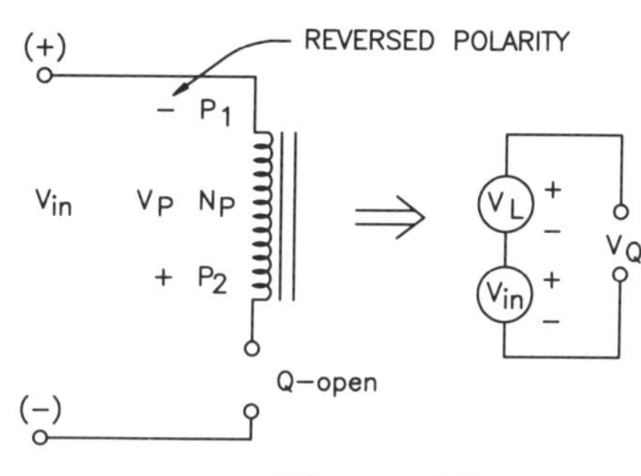

Figure 4.2e

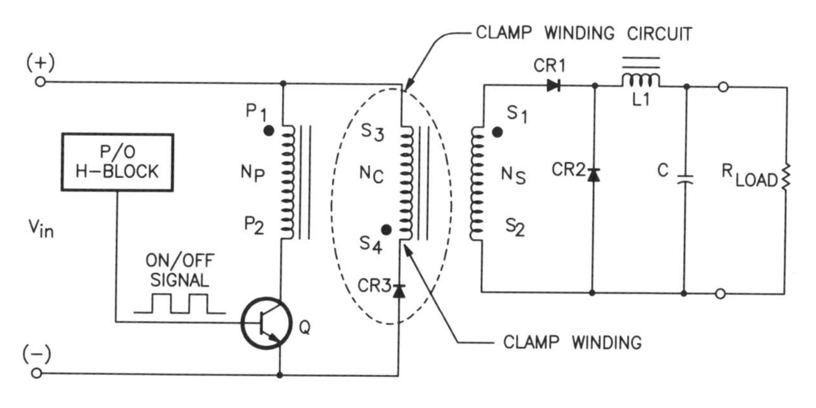

Figure 4.2f

OFF State with Clamp Winding

Unlike the ON state, the clamp winding circuit does have an effect in the OFF state. In this state, when P_1 (dot) becomes negative, S_4 becomes negative and causes CR3 to be forward-biased. (The equivalent circuit is shown in *Figure 4.2h*.) The voltage on the primary coil is coupled to the clamp winding. The clamp winding creates a path to V_{in}, and the energy stored in the inductor will be removed and returned to the input source, V_{in}, returning part of the energy back to the source.

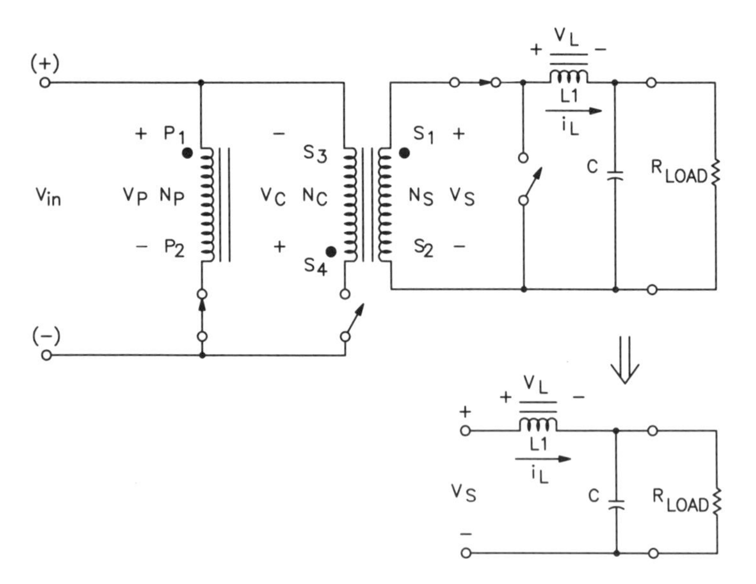

Figure 4.2g

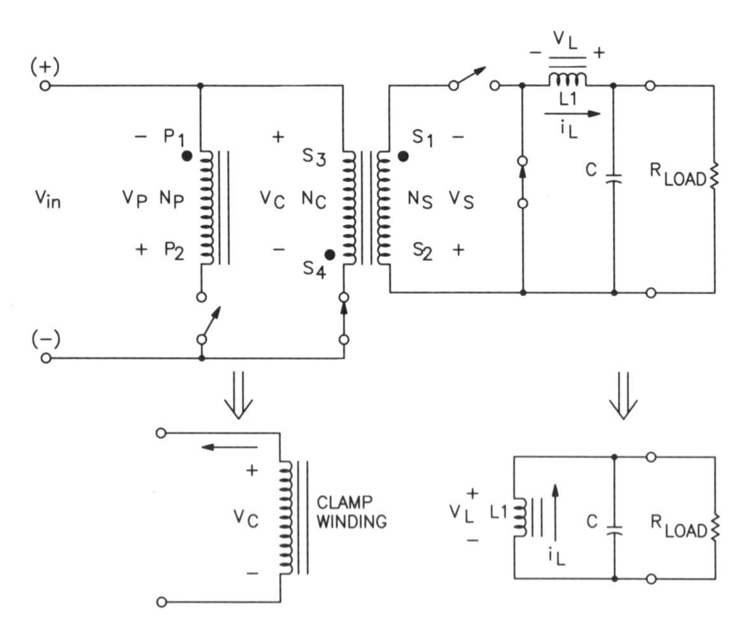

Figure 4.2h

Since the forward converter is an isolated version of the Buck regulator, the mathematical equation of section 3.3.1 ($V_O = D * V_{in}$) can now be modified to reflect the turns ratio of the transformer; therefore, the expression for the forward converter is:

$$Vo = \frac{Ns}{Np} * D * Vin$$

Recall that D is the ratio of t_{on} to T ($D = \frac{ton}{T}$).

4.2.2 FLYBACK CONVERTER

The flyback converter is an isolated version of the Buck-Boost regulator, discussed in section 3.3.3. An ideal circuit for this topology is shown in *Figure 4.3a*. Recall that in the Buck-Boost regulator (see *Figure 3.10*), there are four main components; Q, CR, L, and C. As in the forward converter, all four components remain in the circuit, and an extra component, T_1, is added in the circuit of *Figure 3.10*. In the new configuration, the location of Q was changed and the direction of the CR1 is reversed. Also, notice the positions of the dots on T_1. Unlike the forward converter, these dots are in the opposite directions. T_1 is providing the isolation between the input and the output.

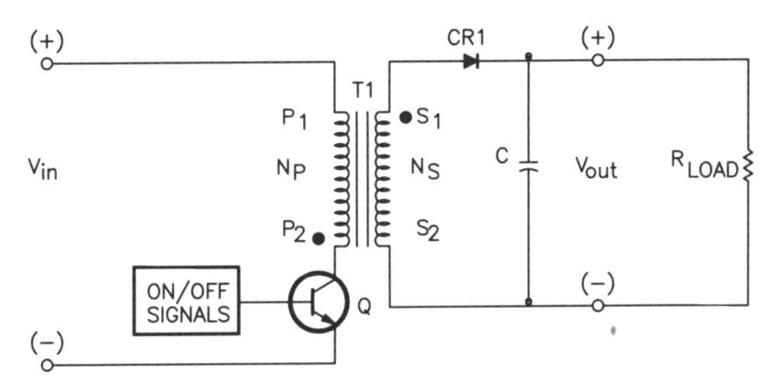

Figure 4.3a

ON State

When Q in *Figure 4.3a* is turned ON, the voltage, V_{in}, appears across the primary winding, N_P. This will cause the current to flow through the coil. The current, i_p, flowing through the primary winding, will cause the P_1 terminal to be positive and P_2 (dot) to be negative. When P_2 (dot) is negative,

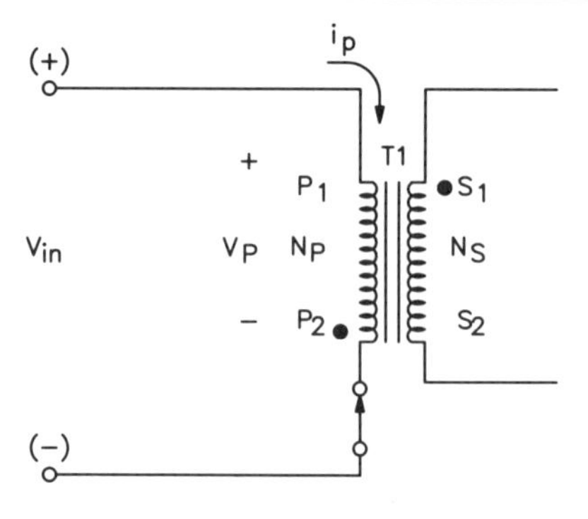

Figure 4.3b

S_1 (dot) will also be negative. The negative polarity on S_1 will cause CR1 to be reversed-biased, and it will act as an open. (The equivalent circuit is shown in *Figure 4.3b*.) In this state, the energy is stored in the primary coil.

OFF State

When Q is turned OFF, the magnetic field across the primary collapses and the polarity of the voltage across the coil reverses. As a result, the P_1 terminal becomes negative and the P_2 (dot) terminal becomes positive; therefore, the S_1 (dot) terminal becomes positive and CR1 is now forward-biased and acts as a short. (The equivalent circuit is shown in *Figure 4.3c*.) The energy stored in the primary coil now is transferred to the secondary. Since the flyback converter is an isolated version of the Buck-Boost, the mathematical expression is:

$$-V_0 = D * \frac{1}{1-D} * V_{in}$$

We can modify this to:

$$V_0 = \left[\frac{Ns}{Np}\right] * D * \frac{1}{1-D} * V_{in}$$

4.2.3 PUSH-PULL CONVERTER

This converter can be seen as two forward converters, and it is called a Buck derived topology. In this topology, there are two switches, Q1 and Q2, as shown in *Figure 4.4a*. Notice that T1 is center tapped on the primary side. For this topology, we will also talk about the ON and OFF states, but

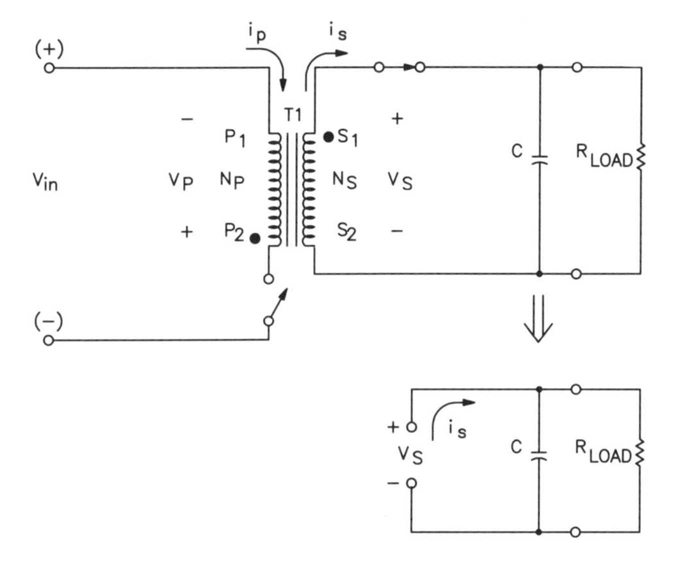

Figure 4.3c

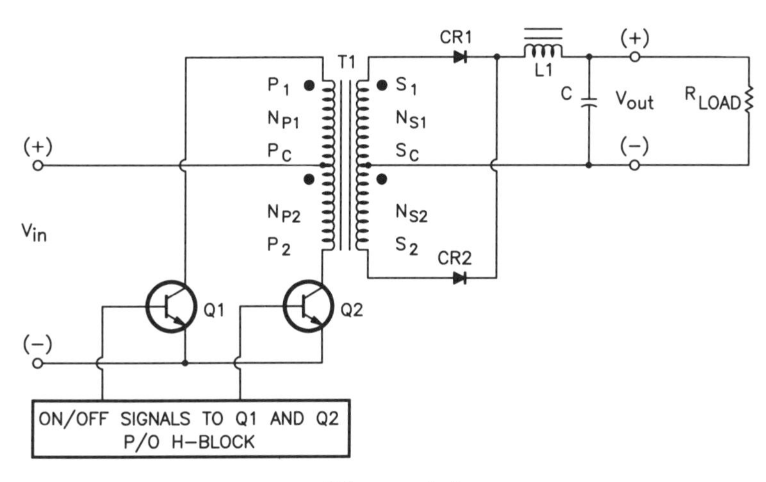

Figure 4.4a

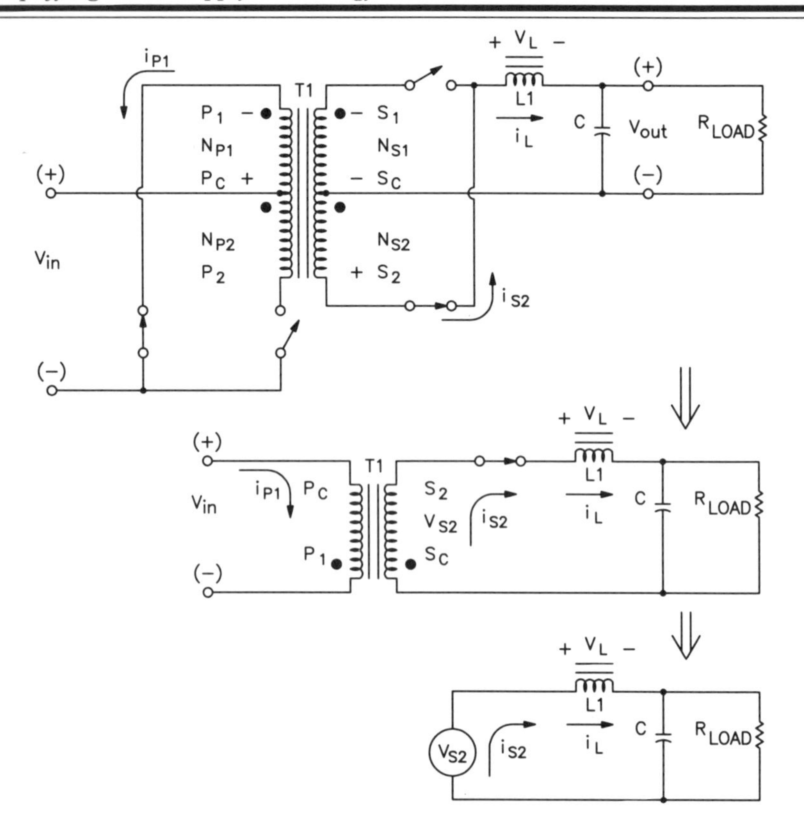

Figure 4.4b

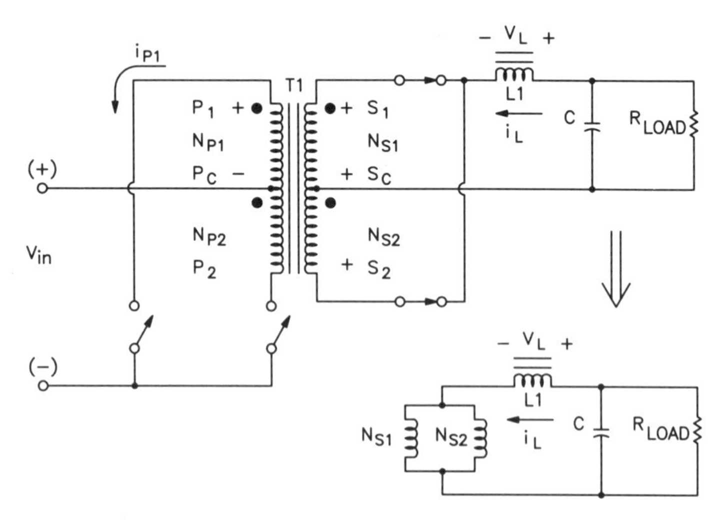

Figure 4.4c

in a slightly different manner because there are four combinations of switch operations that are involved, as opposed to only two in the previous topologies. The states in sequence are:

Q1-ON and Q2-OFF
Q1-OFF and Q2-OFF
Q1-OFF and Q2-ON
Q1-OFF and Q2-OFF

Notice that a state of Q1-ON and Q2-ON is not listed because it is not part of the normal operation of this topology. Interested readers may go through the circuit analysis of this state and see the effect.

Q1-ON and Q2-OFF State

When Q1 turns ON, the current, i_{P1}, will start flowing from V_{in} into P_C and out of P_1 (dot). The polarity at P_C is positive with respect to P_1 (dot); therefore, on the secondary side, S_2 is positive with respect to S_C (and S_1). The positive polarity on S_2 will cause CR2 to be forward-biased. CR2 acts as a short, and CR1 acts as an open. (The equivalent circuit of this mode is shown in *Figure 4.4b.)* The circuit of *Figure 4.4b* is similar to the forward converter ON-state circuit (see *Figure 4.2b*). Refer to section 4.2.2 for details.

Q1-OFF and Q2-OFF State

After some time (t_{on}), Q1 turns OFF and Q2 remains OFF. When both Q1 and Q2 are in OFF states, it is called *dead time.* During this time, the polarity across the N_{P1} winding will reverse, making P_1 (dot) positive; therefore making S_1 positive and causing CR1 to be forward-biased. Also during this time, the voltage across L_1 is reversed, and the current from L_1 will flow into both halves of the secondary side. (The equivalent circuit of this is shown in *Figure 4.4c*.)

Q1-OFF and Q2-ON State

After the dead time is completed, Q2 turns ON and Q1 remains OFF. When Q2 turns ON, it causes the current, i_{P2}, to flow from the V_{in} (+) terminal into the P_C (dot) terminal, and out of the P_2 terminal and back to V_{in}(-). The polarity at P_C (dot) is positive; therefore, polarity at S_1 (dot) is also positive. This will keep CR1 in a forward-biased mode. The voltage polarity at S_2 is negative; therefore, CR2 will be reversed-biased and the bottom half of the secondary will not conduct. (The equivalent circuit of this mode is shown in *Figure 4.4d.* This circuit is also similar to the forward converter ON-

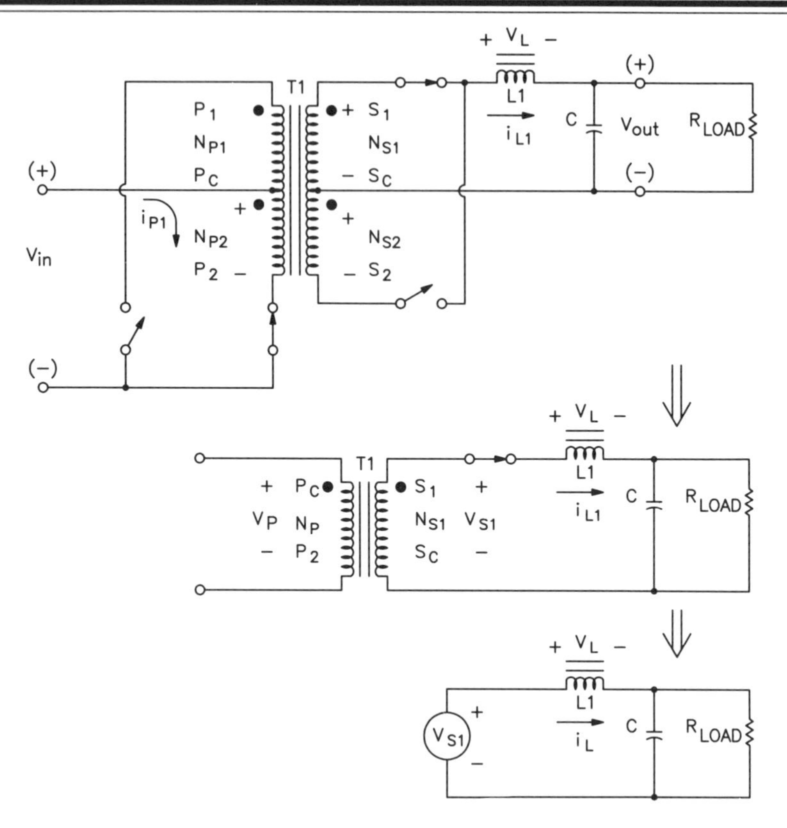

Figure 4.4d

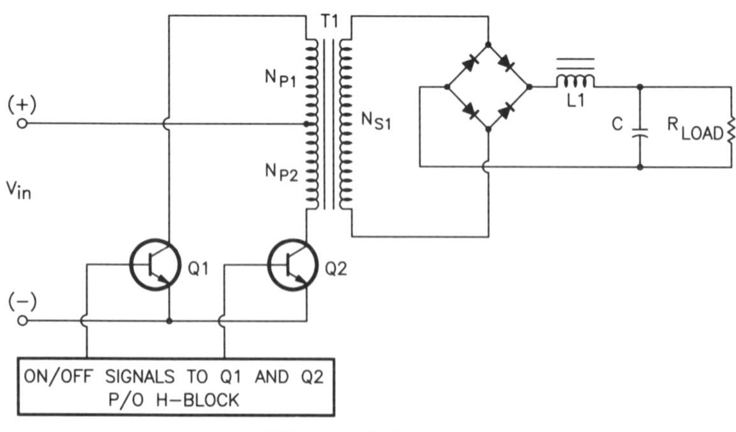

Figure 4.4e

state circuit.) Since Q1-ON/Q2-OFF and Q1-OFF/Q2-ON circuits are similar to the forward converter ON state, we can say that the push-pull topology consists of two forward converters.

Q1-OFF and Q2-OFF State

This state is similar to Q2-OFF and Q1-OFF state. For the sake of completing one cycle, this state is shown.

The circuit in the secondary can be replaced by a full-wave bridge, as shown in *Figure 4.4e*. This configuration will eliminate the need of a center tap on the secondary side.

Mathematical Expression

Since both the primary and secondary coils are center tapped, we say that $N_{S1} = N_{S2}$, and $N_{P1} = N_{P2}$. Therefore, the equation for the push-pull topology is:

$$V_0 = 2 * (N_{S1}/N_{P1}) * D * V_{in}$$

or

$$V_0 = 2 * (N_{S2}/N_{P2}) * D * V_{in}$$

4.2.4 HALF-BRIDGE CONVERTER

This topology also uses two switches, Q1 and Q2, as shown in *Figure 4.5a*, but unlike the push-pull topology, it does not have a center tap on the primary side of the transformer. One major difference in the half-bridge is that there are two capacitors on the input side. Both capacitors (C_1 and C_2) are in series; therefore, the voltage drop on each capacitor is one-half of the input voltage; however, the current in the primary side will be doubled. An ideal circuit is shown in *Figure 4.5a*. In the half-bridge, Q1 and Q2 are turned ON and OFF in the same manner as in the push-pull topology.

Q1-ON , Q2-OFF State

When Q1 turns ON, the current, i_p, flows into the P_1 (dot) terminal and leaves the P_2 terminal; therefore, P_1 (dot) is positive polarity and S_1 (dot) is more positive than S_C or S_2. Positive polarity on S_1 will cause CR1 to be in forward-biased mode and CR2 to be in reversed-biased mode. (The equivalent circuit for this mode is shown in *Figure 4.5b*. In *Figure 4.5b*,

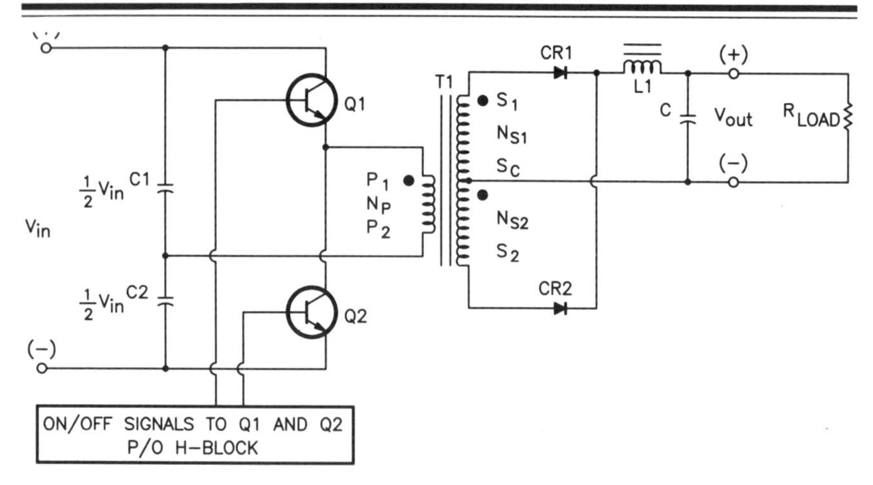

Figure 4.5a

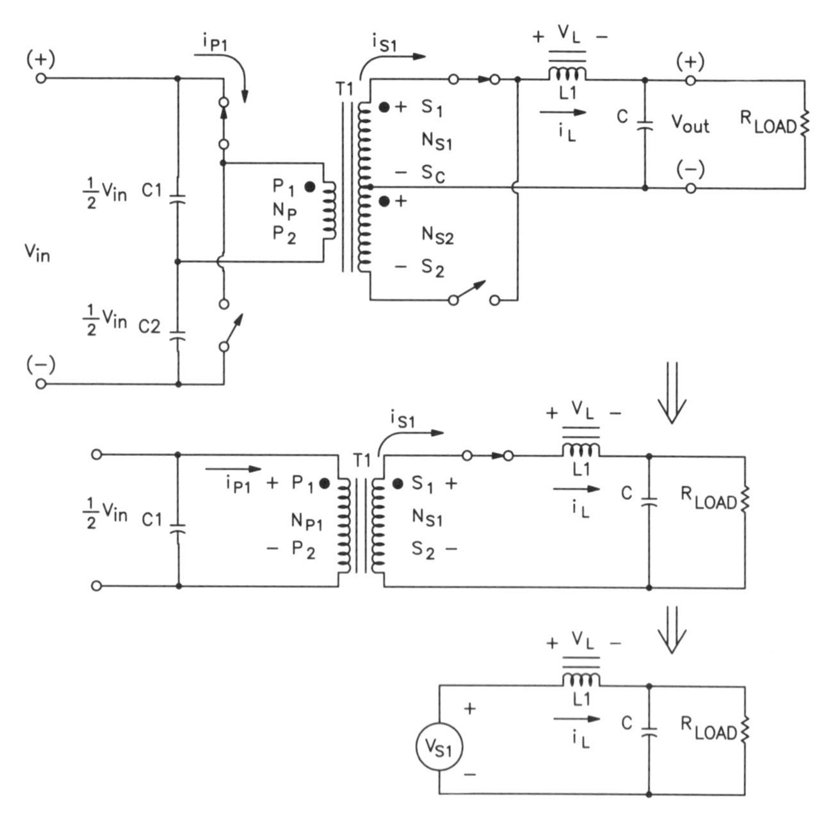

Figure 4.5b

C_2 is out of the circuit and only one-half of the voltage appears across the primary, N_P. This circuit also appears as a forward converter. Refer to section 4.2.2 for a detailed explanation of this circuit).

Q1-OFF , Q2-OFF State

In this state, when Q1 goes from ON to OFF, the voltage polarity across N_P reverses and P_1 (dot) becomes more negative than P_2; therefore, S_1 (dot) becomes more negative than S_C and reverse-biases CR1. However, as soon as CR1 acts as an open, the current through L_1 stops flowing in the direction of the load; therefore, the voltage across L_1 reverses the polarity and current starts to flow in the opposite direction. The negative voltage polarity at the CR1 cathode forces it into a forward-biased condition, and the inductor current flows through CR1, decaying the current through L_1. This is the same operation as in the push-pull topology (see *Figure 4.5c*).

Q1-OFF , Q2-ON State

In this state, when Q2 turns ON, the current, i_{P2}, flows from C_2 into P_2, making the P_2 terminal more positive than the P_1 (dot) terminal. Since the P_1 (dot) terminal is negative, the terminals S_C and S_1 will become more negative than S_2, causing CR1 to be reversed-biased; therefore, it acts as an open. Since the S_2 terminal is now positive, CR2 will become forward-biased. (The equivalent circuit for this mode is shown in *Figure 4.5d*. As seen from *Figure 4.5d,* the final circuit once again looks like the circuit of a forward converter. See section 4.2.2 for a detailed operation.)

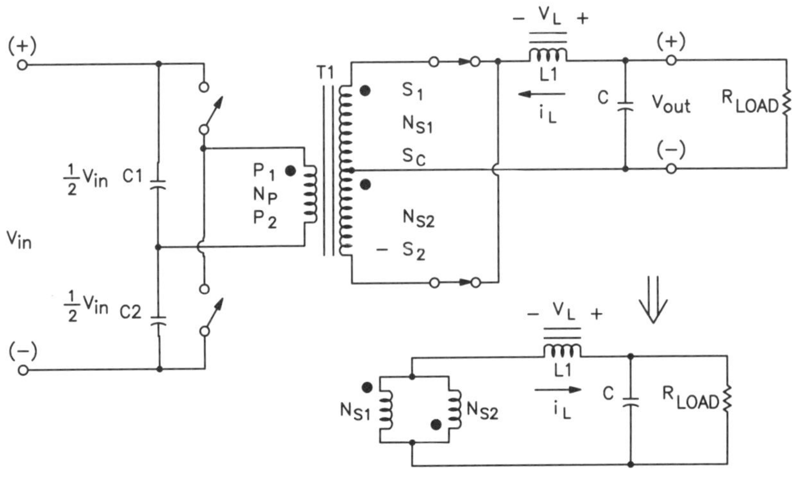

Figure 4.5c

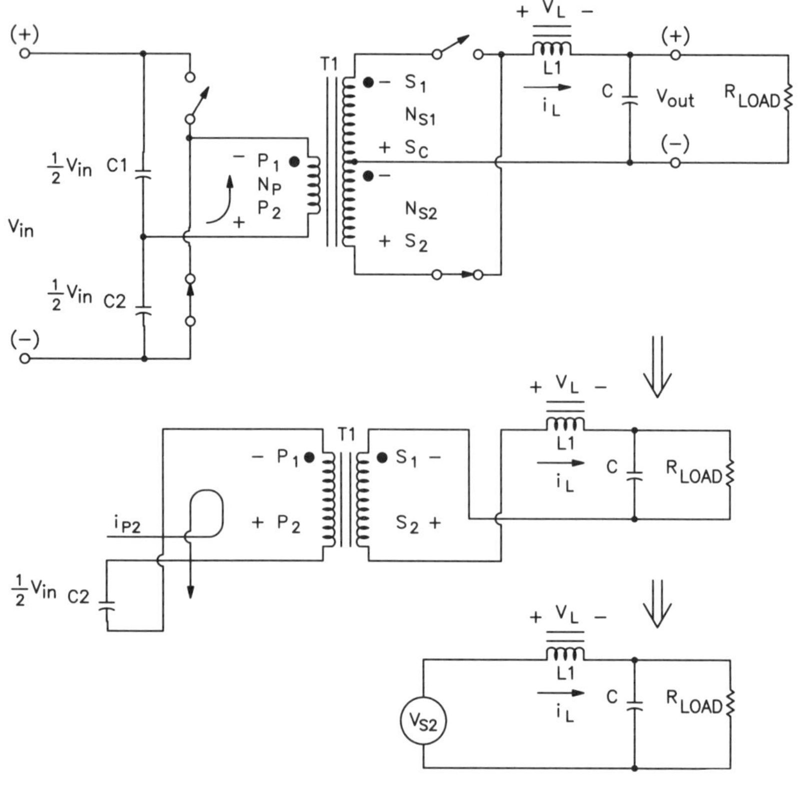

Figure 4.5d

Q1-OFF, Q2-OFF State

This state is similar to the Q1-OFF and Q2-OFF state described earlier. For the sake of completing one cycle, this state is shown.

Mathematical Expression

Since the secondary coil is center tapped, we say that $N_{S1} = N_{S2}$. Therefore, the equation for the half-bridge topology is:

$$V_0 = (N_{S1}/N_P) * D * V_{in}$$

or

$$V_0 = (N_{S2}/N_P) * D * V_{in}$$

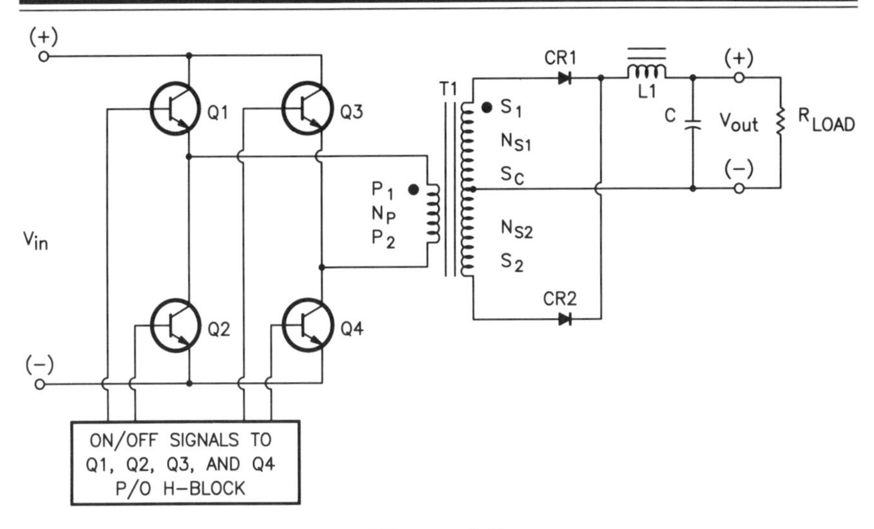

Figure 4.6a

4.2.5 FULL -BRIDGE CONVERTER

Of all the topologies that we've examined so far, the full-bridge converter topology is the only one that uses four switches; Q1, Q2, Q3 and Q4. The ideal circuit is shown in *Figure 4.6a*. Unlike the half-bridge, this topology has only one input capacitor. Although it seems that there are many switching state combinations, only four are commonly used. They are:

Q1 & Q4-ON , Q2 & Q3-OFF
Q1 & Q4-OFF , Q2 & Q3-OFF
Q1 & Q4-OFF, Q2 & Q3-ON
Q1 & Q4-OFF , Q2 & Q3-OFF

Notice that Q1 and Q4 always turn ON and OFF at the same time, and Q2 and Q3 turn ON and OFF at the same time. When these combinations occur, the circuit looks similar to a half-bridge topology, except the voltage across the primary coil is full V_{in}, rather than one-half V_{in} in the half-bridge. Therefore, the current is one-half the half-bridge (assuming the power delivered by both topologies is the same).

Q1 & Q4-ON, Q2 & Q3-OFF State

The switches Q1 and Q4 are turned ON at the same time, causing the current, i_{P1}, to flow through the primary coil, N_P. Since the current is entering the P_1 (dot) terminal, the voltage polarity will be positive. Therefore, S_1 (dot) will be more positive than S_C or S_2. (The equivalent circuit of this state is shown in *Figure 4.6b.*) Once again, the equivalent circuit of this state is reduced to the circuit of the forward converter.

Q1 & Q4-OFF, Q2 & Q3-OFF State

When Q1 and Q4 turn OFF, and Q2 and Q3 remain OFF, the voltage in the primary, N_P, reverses the voltage polarity. The P_1 (dot) terminal now becomes negative, S_1 becomes more negative, and CR1 acts as an open. However, the collapsing magnetic field on L_1 reverses the voltage polarity and forward-biases CR1, creating a current path for the L_1. This will cause the current through L to decay. (The equivalent circuit in the dead time region is shown in *Figure 4.6c.*)

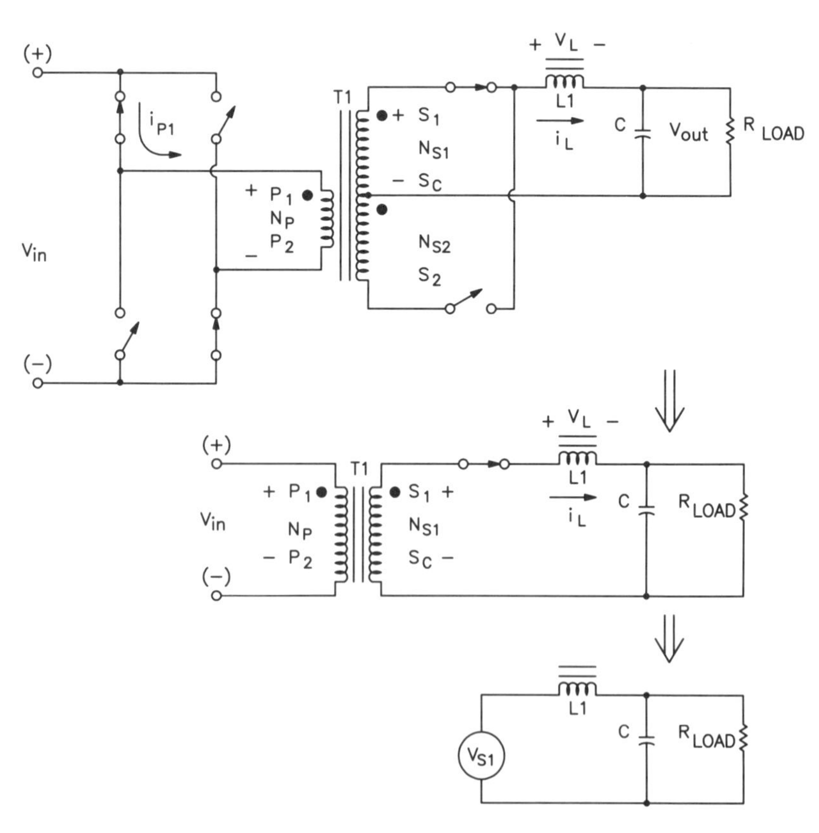

Figure 4.6b

Q1 & Q4-OFF, Q2 & Q3-ON State

When Q2 and Q3 turn ON, the current, i_{P2}, starts flowing in the opposite direction through the primary coil, N_P. The current enters the P_2 terminal and leaves the P_1 (dot) terminal. The P_1 terminal is positive and P_2 (dot) is negative; therefore, S_2 (dot) is negative and CR1 is open. S_1 is positive and CR2 becomes forward-biased, creating a short, as shown in *Figure 4.6d.*

Q1 & Q4-OFF, Q2 & Q3-OFF State

This state is the same as Q2 and Q3-OFF, and Q1 and Q4-OFF. For the sake of completing one cycle, this state is shown.

Mathematical Expression

Since the secondary coil is center tapped, we say that $N_{S1} = N_{S2}$. Therefore, the equation for the half-bridge topology is:

$$V_0 = 2 * (N_{S1}/N_P) * D * V_{in}$$

or

$$V_0 = 2 * (N_{S2}/N_P) * D * V_{in}$$

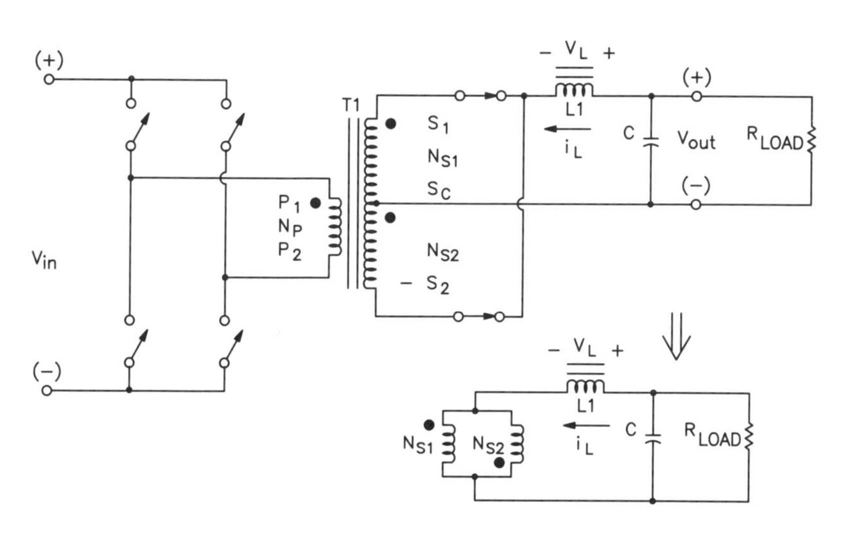

Figure 4.6c

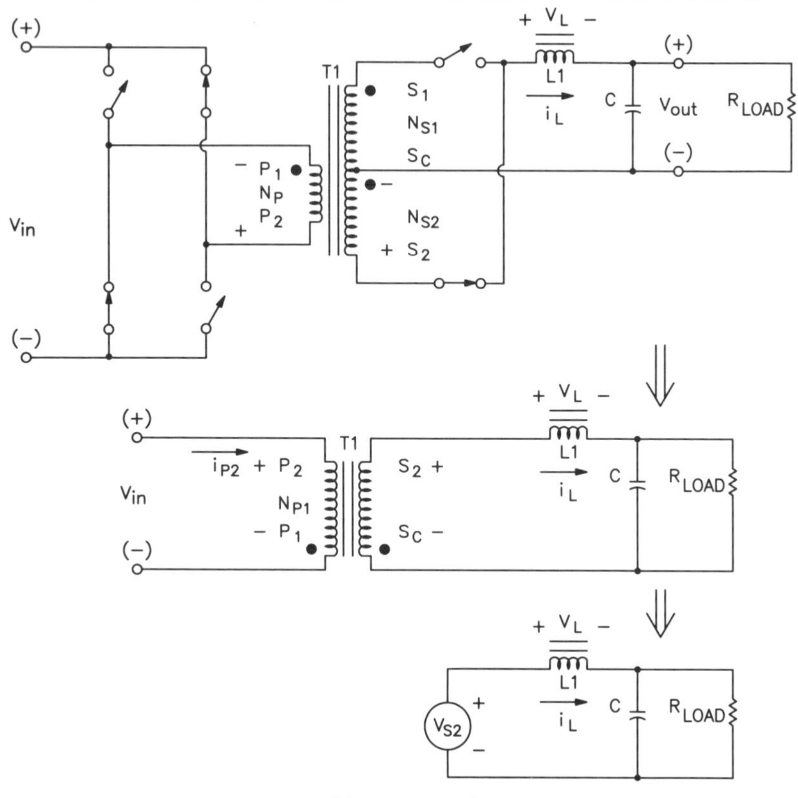

Figure 4.6d

4.2.6 LIMITATIONS OF PWM TECHNIQUE

The above converters use PWM technique. Although this is a popular method, it has some drawbacks. To understand these drawbacks, we will have to closely examine the operation of the transition states of the switching device, Q. There are two transition periods; ON transition and OFF transition. In the ON transition, the output is turning ON, meaning that the voltage across the collector to the emitter of the transistor is transitioning from V_{CC} to 0, and the collector current is transitioning zero to the peak current. In the OFF transition, the collector is turning OFF, meaning that the voltage across the collector to the emitter is transitioning from 0 to V_{CC}, and the collector current is transitioning from peak current to 0. Since we have been using a bipolar (NPN) transistor as a switching device, we will continue with the same device to understand the basic concepts in this section; however, other devices such as *field effect transistors* (FETs), etc., can be substituted for Q. First, we will examine the switching device in a resistive circuit, then in the inductive circuit.

Understanding the Energy Loss in a Switching Device

Switching Device in the Resistive Circuit

Let's examine the transistor, Q, in the resistive circuit, as shown in *Figure 4.7a*. When a positive voltage is applied at the V_B terminal, the base current, I_B, starts flowing into the base circuit. Assuming that this current is sufficient to turn the transistor ON, the collector current, I_C, will start flowing into the "collector circuit." Where I_C depends on the value of I_B and the transistor gain, β, or $I_C = \beta * I_B$.

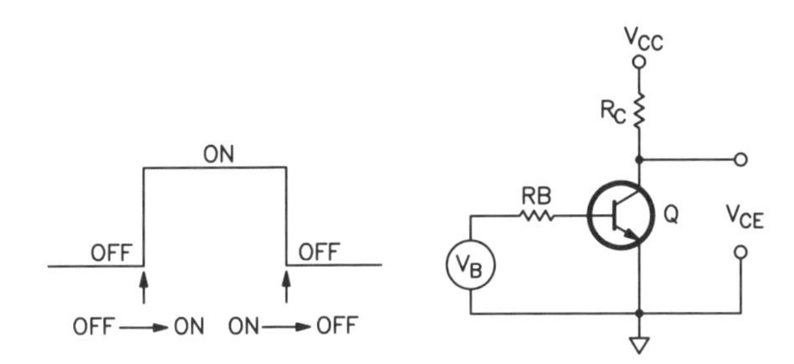

Figure 4.7a

I_C	$V_{CC} - I_C * RC = V_{CE}$	$V_{CE} * I_C =$ Power Loss
12	24 - (12 * 2) = 0	0 * 12 = 0 Watts
11	24 - (11 * 2) = 2	2 * 11 = 22 Watts
10	24 - (10 * 2) = 4	4 * 10 = 40 Watts
9	24 - (9 * 2) = 6	6 * 9 = 54 Watts
8	24 - (8 * 2) = 8	8 * 8 = 64 Watts
7	24 - (7 * 2) = 10	10 * 7 = 70 Watts
6	24 - 6 * 2) = 12	12 * 6 = 72 Watts
5	24 - (5 * 2) = 14	14 * 5 = 70 Watts
4	24 - (4 * 2) = 16	16 * 4 = 64 Watts
3	24 - (3 * 2) = 18	18 * 3 = 54 Watts
2	24 - (2 * 2) = 20	20 * 2 = 40 Watts
1	24 - (1 * 2) = 22	22 * 1 = 22 Watts
0	24 - (12 * 2) = 0	24 * 0 = 0 Watts

Table 4.1

If we assume that β remains constant, then I_C is strictly controlled by I_B. If we pass sufficiently large I_B while making sure that the transistor is not damaged, I_C will also be large. When I_C becomes larger, V_{CE} will become smaller, as shown by the following equations. Using Kirchoff's Voltage Law:

$$V_{CC} = V_{RC} + V_{CE}$$

$$V_{CC} = (I_C * R_C) + V_{CE}$$

$$V_{CE} = V_{CC} - (I_C * R_C)$$

If I_C becomes larger and larger, then the $I_C * R_C$ term will become larger and the difference will become smaller. Eventually, $I_C * R_C$ will reach a value of V_{CC}; therefore, $V_{CC} - I_C * R_C = 0$, or $V_{CE} = 0$. I_C will continue to flow as long as I_B is flowing into the base. However, as soon as the base current is removed, I_C must stop flowing immediately and go down to zero (OFF transition). But that is not the case in the practical transistor. In the OFF-transition, the output or collector current decreases gradually. If we look at the above equations again, we will see that as I_C decreases, the voltage across R_C decreases, and V_{CC} - $I_C * R_C$, or V_{CE} increases. In the next example, we will see how this gradual transition causes power dissipation in the switching device.

Let us assign some values to the components in *Figure 4.7a* to closely examine the effect. Let's say that the collector current was 12A, $R_C = 2\Omega$ and $V_{CC} = 24V_{DC}$. (*Table 4.1*) The values in *Table 4.1* are plotted in *Figure 4.7b*. The pulse in *Figure 4.7b* represents the amount of power dissipated in the

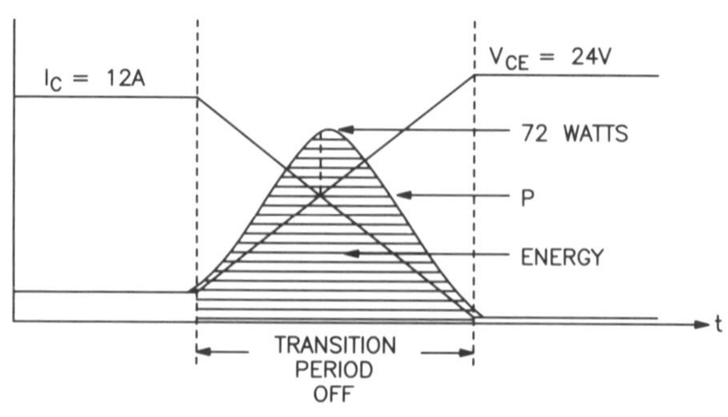

Figure 4.7b

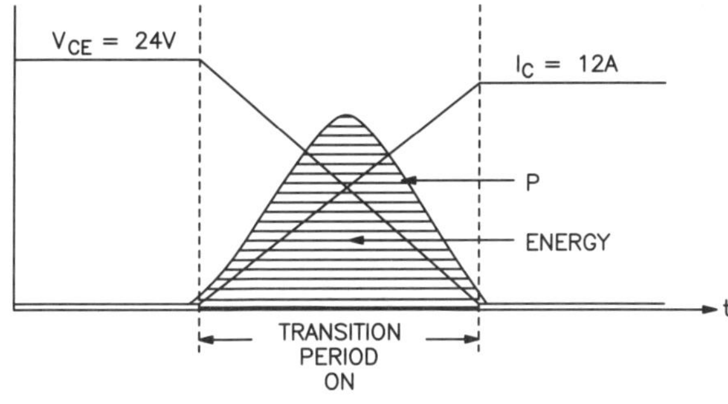

Figure 4.7c

switching device during the OFF-transition period. This loss is commonly defined as energy loss. Recall that Energy = Power * Time, where energy is in joules, power in watts, time in seconds

Such pulse also exists when V_B goes from 0V to 5V. This is the ON-transition (see *Figure 4.7c*). The energy (or power) was dissipated in the switching device during the ON and OFF switching transitions; therefore, it is called a *switching loss.*

There is another type of energy loss that is of interest. In most practical transistors, V_{CE} does not actually go down to 0V in the ON-state. It goes to a level called V_{CE}-SAT, which is specified by the manufacturer. It is called V_{CE}-SAT because it stands for the voltage across the collector to emitter during saturation (or when the device is fully turned ON). When the device is fully ON, the maximum current flows through the collector; therefore, the product of V_{CE}-SAT and the collector current is the power dissipation during the conduction period. This can also be represented in the energy loss, as described in the previous discussion. This power was dissipated when the device was conducting, as opposed to when it was transitioning the level; therefore, it is called a *conduction loss*, as shown in *Figure 4.7d.*

Now suppose that V_B is a train of voltage pulses, as shown in *Figure 4.7e.* I_B and I_C will also see a similar pulse train; therefore, in our example, I_C will transit between the +12A and 0A level, and V_{CE} will transit between +24V and V_{CE}-SAT. Both V_{CE} and I_C will transit this level gradually. If we overlap these two curves as we did in the previous example, we will see that for each transition there is a power (or energy) pulse similar to the ones in *Figures 4.7b* and *4.7c*; and during each conduction period, the power

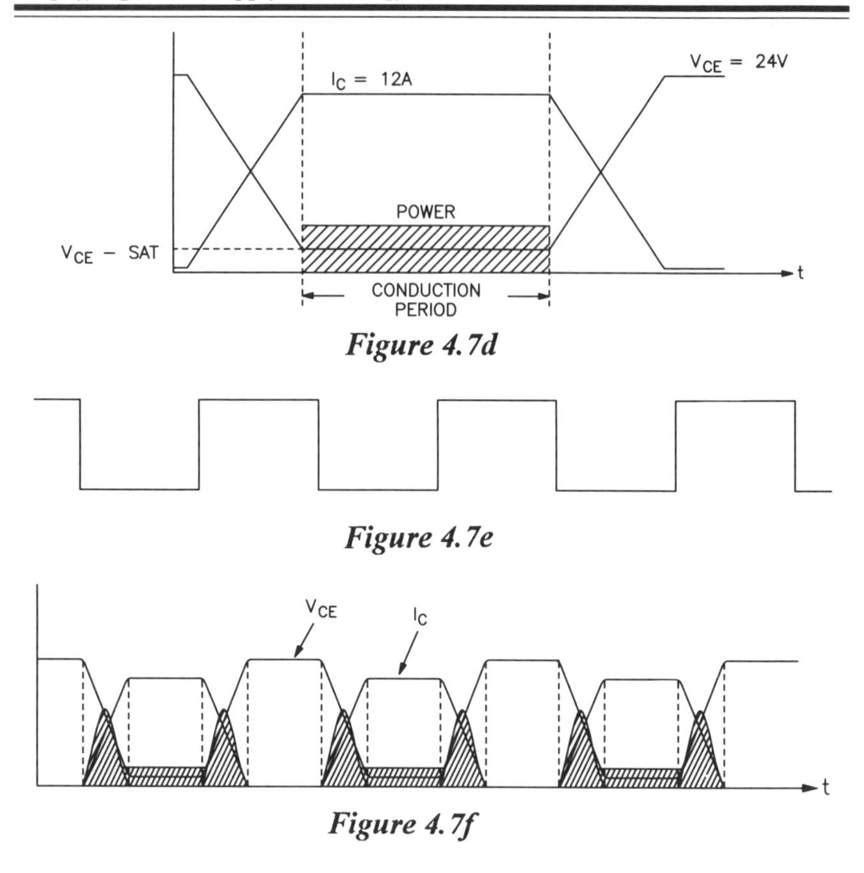

Figure 4.7d

Figure 4.7e

Figure 4.7f

pulse is similar to *Figure 4.7d.* In the train of pulses, the power is dissipated during each transition period and each conduction period, as shown in *Figure 4.7f.*

The total power loss in the switching device can be expressed as follows:

$$P_{device} = P_{switching} + P_{conduction}$$

Switching Device in the Inductive Circuit

Since all of our converter circuits use a transformer in the collector circuit of Q, let us examine the effect it has in the voltage and current of the switching device. We can replace the transformer with an inductor.

We will first look at a *resistor-inductor* (R-L) circuit to understand the voltage current relations (see *Figure 4.8a*). Let's say that at t_0, SW1 closes. The current from the battery will start flowing into the base (thus collector) circuit.

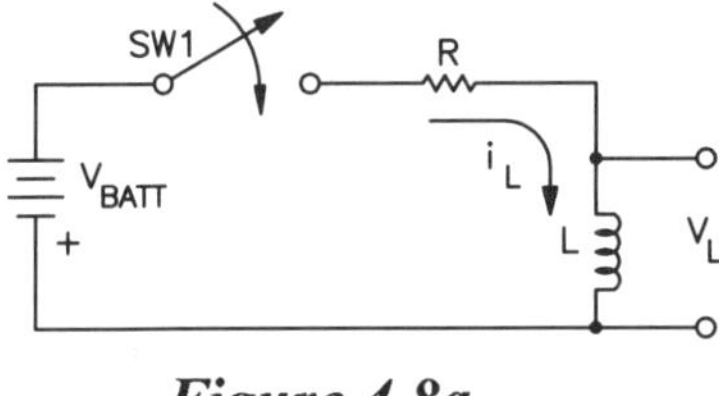

Figure 4.8a

In the DC circuit, initially, the inductor acts as an open circuit, and after some time it acts as a short circuit. During the time interval—when it is open to when it is short—the inductor is said to be charging up. The time it takes to fully charge is "five times the constant," where the constant is the ratio of L and R:

τ = L/R, where t is in seconds, L is in Henries, and R is in Ohms

The value of τ is a one time constant. For the inductor to fully charge, it takes 5 * τ. The voltage across an inductor is given by $V_L = V_{battery} * e^{-t/t}$. And the current through the inductor is given by:

$$i_L = i_{FINAL}*(1 - e^{-t/t}), \text{ where } i_{FINAL} = \frac{V_{batt}}{R}.$$

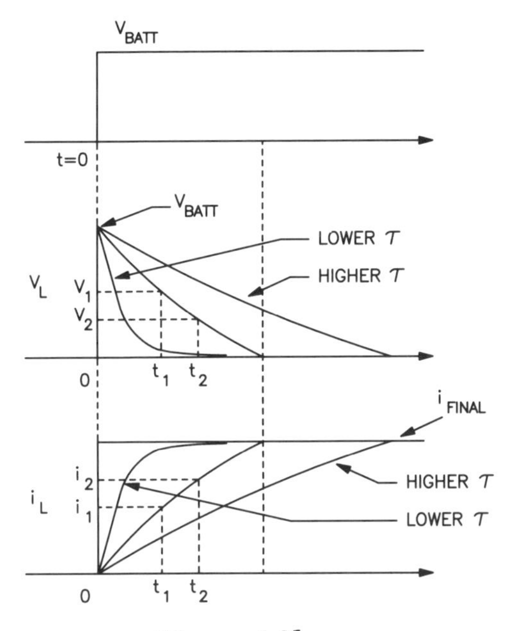

Figure 4.8b

If we were to tabulate different values of V_L and i_L, as a function of L and R, we would see the general response, as shown in *Figure 4.8b*. When the values of R and L change, the slopes of V_L and i_L change. For example, a circuit with the τ of 4µs would have a curve with a steeper slope than the one with the τ of 10µs; however, V_L and i_L reach the same final values in both instances. In our discussion, the slope is the difference in two values of currents or voltages divided by the difference in the corresponding time values. For example:

$$\frac{V2-V1}{t2-t1} = \frac{4V-6V}{2\mu s-1\mu s} = \frac{-2V}{1\mu s} = -2V/\mu s$$

$$\frac{L2-L1}{t2-t1} = \frac{8mA-6mA}{2\mu s-1\mu s} = \frac{2mA}{1\mu s} = 2000A/\mu s$$

Another way of writing V_L is by using the slope method, or:

$V_L = L * \frac{di}{dt}$, or L times the slope of the current through the inductor.

Now let's substitute L in place of R_C in *Figure 4.7a* and examine this circuit (see *Figure 4.9a*). When V_B goes high (+5V), I_B starts flowing into the base circuit and I_V starts flowing through the inductor, L. This current will rise gradually; therefore, it will have a slope and V_L can be calculated. If we turn V_B off, I_B stops flowing, causing I_C to stop; but as soon as I_C stops, the voltage across L will reverse voltage polarity. The voltage across V_{CE} is now V_{CC} - V_L = V_{CE}, but since V_L has reversed its polarity, the above expression will be V_{CC} - (-V_L) = V_{CE}; or rewriting the above equation, $V_{CE} = V_{CC} + V_L$.

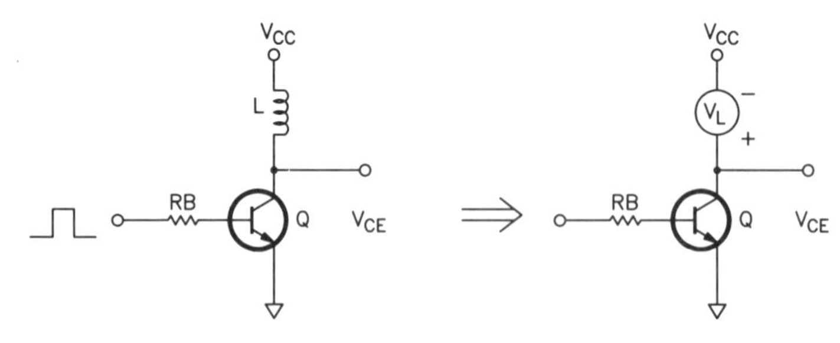

Figure 4.9a

Notice the + sign in the expression. The total voltage across the collector emitter (V_{CE}) will be higher than the V_{CC}. The higher voltage is the voltage across the inductor (V_L), and this voltage is called a *voltage overshoot.* After it reaches this $V_{CE}pk$ value, and assuming that there is a discharge path for the energy in L, V_{CE} will come down to V_{CC} level, as shown in *Figure 4.9b*. While V_{CE} and I_C are transitioning the levels, the switching power is dissipated. Also, due to the voltage overshoot, the switching device is seeing more stress on the collector to emitter.

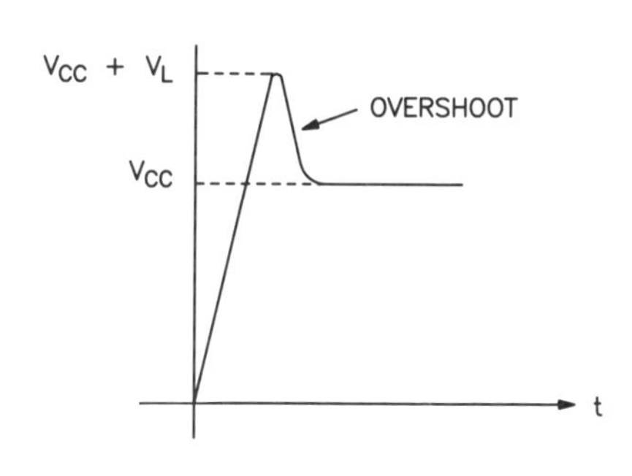

Figure 4.9b

If the voltage overshoot becomes too high, then it may exceed the maximum voltage limit on Q and may cause the device to degrade or fail. This is also known as exceeding the *safe operating area*, or SOA. It is the region specified by the manufacturer. From the SOA curve, one can see the maximum combinations of the voltage and current that the device can operate safely.

Snubber Circuits

In our conductive circuit there are some limitations. In order to protect the switching device from both the voltage overshoot and the switching losses, a technique of absorbing the energy is used. The circuit that performs this function is called a *snubber circuit.*

Dissipative

The snubber circuit reduces the peak value of the voltage overshoot and reduces the device's switching loss. Although there are a few variations of the snubber circuit, we will examine the operation of the circuit in *Figure 4.10a*. Let's assume that the current, I_C, in the circuit has been flowing for some time, then Q is turned OFF. (The equivalent circuit in this mode is shown in *Figure 4.10b*.) As soon as Q opens, the voltage at the collector terminal will become positive, CR1 will become forward-biased, and capacitor, C, will start charging, initially creating a short across Q; therefore, all the current will flow into C. In this mode, the voltage across the collector to the emitter (V_{CE}) is transitioning from 0 to V_{CC} without any current flowing through it; therefore, there is no switching loss.

When Q is turned ON, it will apply a short (or V_{CE}-SAT) across the snubber circuit, as shown in *Figure 4.10c*. This state will cause the capacitor to discharge through the resistor R. Since the capacitor is discharging through the resistor, all the current will pass through it, and the resistor will dissipate this stored energy. The power dissipation in the resistor = $i^2 * R$. At this point, we have transferred the switching power loss into a resistive power loss. The power loss is still there; we simply transferred it from the transistor to the resistor; and because the power dissipates in the circuit, it is called a dissipative snubber circuit.

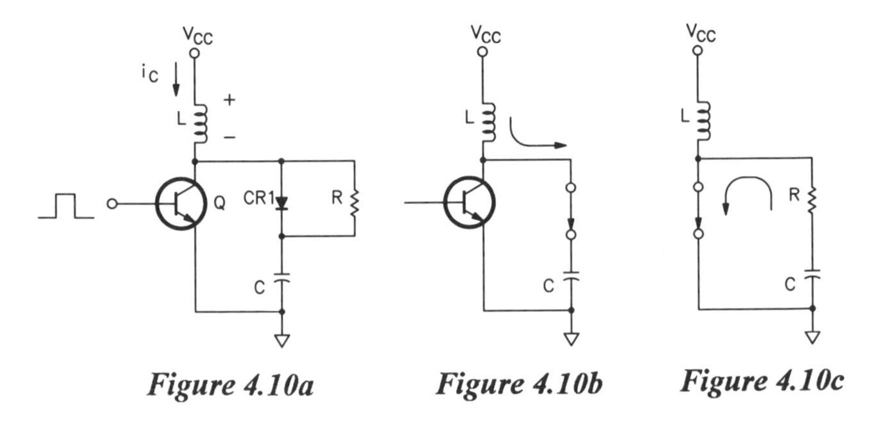

Figure 4.10a ***Figure 4.10b*** ***Figure 4.10c***

Non-Dissipative

Though this circuit dissipates power, it is still a preferred method by many designers; however, non-dissipative snubber circuits have also been developed. This method also absorbs extra energy, but instead of dissipating it, it returns it back into the source, saving energy. Let's examine the operation of this circuit in detail.

To simplify the concept, let's first assume that point A in *Figure 4.11a* is charged to V_{in}, and the current, I_C, has been flowing in the circuit for some time. When Q turns OFF, the voltage across L_P will reverse its polarity and cause the current to flow into C_S, through CR1 and back into L_P. (The equivalent circuit for this mode is shown in *Figure 4.11b*.) CR2 is reversed-biased; therefore, no current flows in L_S. In this mode, the current from L_P will charge C_S. When C_S is fully charged, the current will stop flowing in the circuit. In this mode, the energy that was stored into L_P is now transferred into C_S; therefore, no power was dissipated. Simply, a transformation occurred. C_S reverses the voltage polarity across it. When Q is turned ON, the current will flow from C_S through Q, to CR2, to L_S, and back to C_S. (The equivalent circuit for this phase is shown in *Figure 4.11c.*)

Again, in this mode, the energy simply transferred from C_S into L_S and did not dissipate; so the energy that was stored into L_P is now stored into L_S. When L_S is fully charged, the capacitor has discharged and the current through L_S will stop flowing, reversing the voltage polarity. (The equivalent circuit for this state is shown in *Figure 4.11d.*) This will make CR1 and CR2 forward-biased, and the energy of L_S (also known as trapped energy) will now be returned to C_{in} or to the source. Extra energy that was stored in L_P was transferred into C_1, then into L_S, and finally back to the source. This process occurs for every switching cycle. Ideally, there is no power dissipation in the switching device or anywhere else in the circuit!

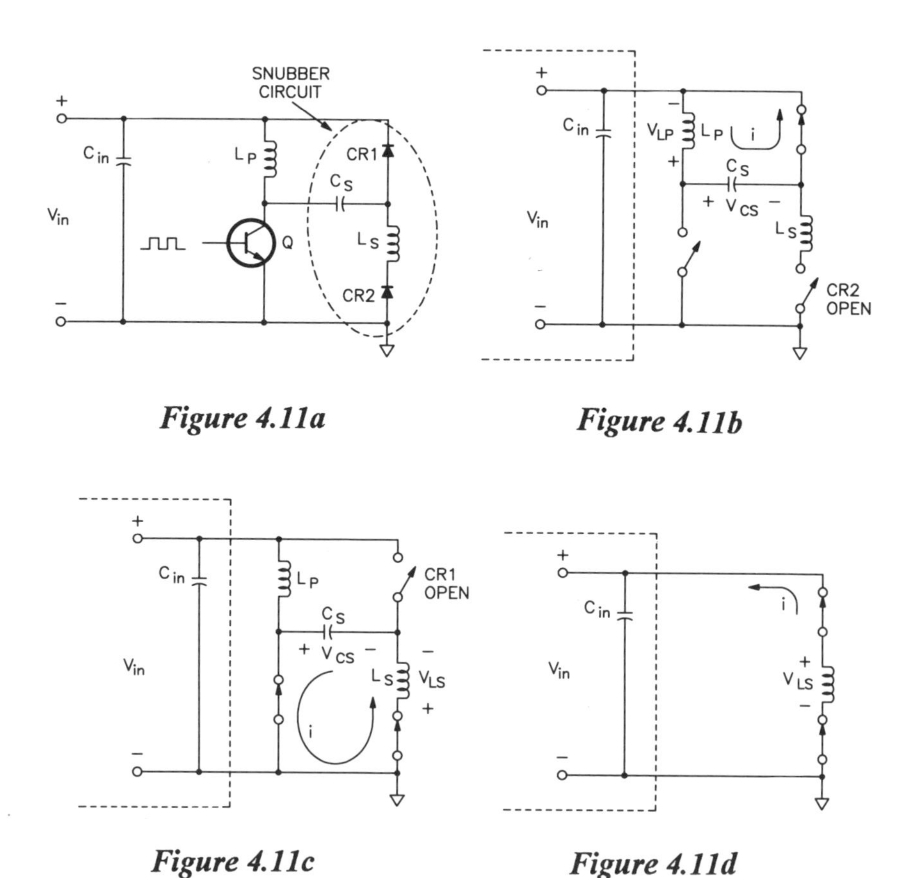

Figure 4.11a

Figure 4.11b

Figure 4.11c

Figure 4.11d

At the present, PWM switching is very popular in the industry, but as the demand for reducing size and weight increases, the need for smaller and lighter components increases. One way to achieve the size and weight requirements is by increasing the switching frequency; but as we saw in the previous sections, increasing the frequency causes higher switching losses and higher voltage stress on the switching devices. It also causes higher EMI noise. To overcome these problems, another set of circuits has been developed called *resonant converters*. Resonant converters operate by keeping the ON and OFF time constant, but this technique changes the frequency, which in turn changes the output voltage. Interested readers may refer to other literature on the subject.

CHAPTER 5
THE CONTROL SECTION

5.1 INTRODUCTION

The purpose of the control section is to make sure that parameters such as the output of a system are maintained at their desirable values. In Chapter 3 we talked about a constant voltage output. If the output voltage changes to an undesirable value, then the control circuit will adjust certain parameters to bring the output voltage back to its original value. Another example is to maintain a constant output current. In this case, the control circuit will maintain a desired current value.

The control section can also protect the system from any catastrophic failures. For example, if an internal temperature of the system exceeds a certain value and creates an abnormal condition, such as the transistor to fail or the inductor to saturate, etc., then it can shut down the system. In case of a failure, the control section can also be designed to inform the user, either by activating certain fault indicator signals or by sending appropriate message signals to the computer or a printing device. So, the control section of the power supply monitors some critical parameters and maintains the desired system operation. In Chapter 3, we also talked about G and H-blocks, where the G-block produces a desired value of V_O, while the H-block monitors to see if the desired value of V_O is maintained. If not, it will send control signals to the G-block to adjust V_O. In this chapter, we will examine the detailed operation of the H-block. First, we will look at some basic concepts and characteristics of a control loop.

The output of the G-block is sensed and fed into the input of the H-block, and the output of the H-block is connected to the G-block. This type of connection creates a feedback loop; it is feeding back information about the output. There are two types of feedback loops; the positive feedback loop and the negative feedback loop. In the control section of power supplies, we talk about the negative feedback loop. This loop is also called a closed loop because the interconnections between the G and H-blocks create a loop that is closed. If the G and H-blocks are not connected as discussed, then it is called an open loop. For example, if V_O is not sensed and fed into the H-block, or if the output of the H-block is not connected to the G-block, or if any of the connections in the path of the loop become open, then the loop is called an open loop. We can understand the behavior of a system by studying the characteristics of the open and closed loops.

5.2 TRANSFER FUNCTIONS

The behavior of a system can be expressed in either a mathematical form or a graphical form; however, it is common practice to express it in a mathematical form. The mathematical expression is called a *transfer function.* In the following sections, we will see how transfer functions are derived and how they are used to understand the general behavior of systems.

5.2.1 BASIC CONCEPT

Example: Resistor Network

For the voltage divider network of *Figure 5.1*, we can write an equation for V_O as follows:

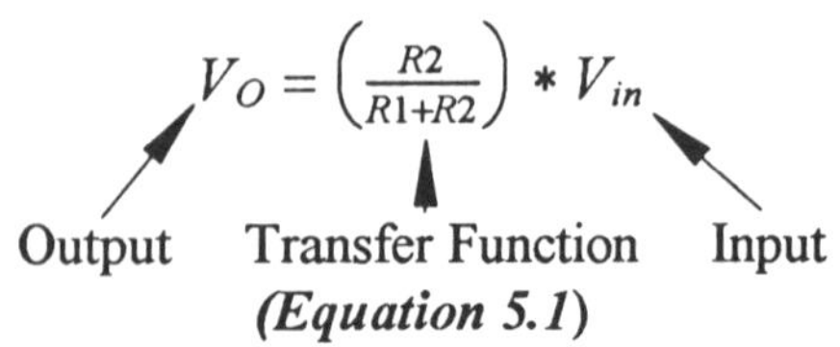

$$V_O = \left(\frac{R2}{R1+R2}\right) * V_{in}$$

(Equation 5.1)

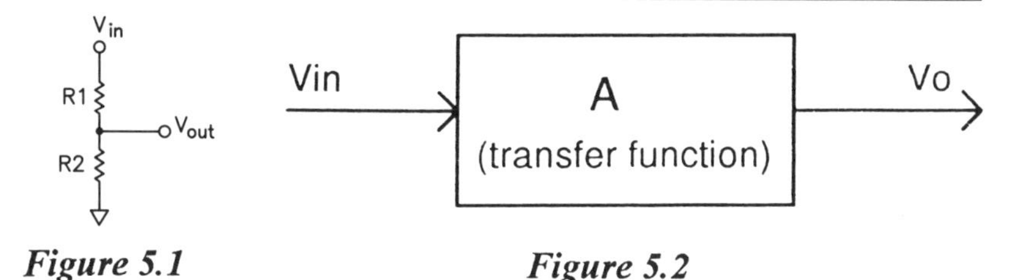

Figure 5.1 ***Figure 5.2***

Dividing both side of *Equation 5.1*, we get:

$$\frac{Vo}{Vin} = \left(\frac{R2}{R1+R2}\right)$$

(Equation 5.2a)

$$\text{Let } A = \frac{Vo}{Vin}$$

(Equation 5.2b)

Substituting *Equation 5.2b* into *Equation 5.2a*:

$$A = \left(\frac{R2}{R1+R2}\right)$$

Transfer Function

(Equation 5.3)

Substituting *Equation 5.3* into *Equation 5.1*:

$$V_O = A * V_{in}$$

(Equation 5.4)

The transfer function (*Equation 5.3*) is also called the *gain* of the network. *Equation 5.4* can be shown in block diagram form as shown in *Figure 5.2*.

Notice that *Equation 5.1* is a generalized equation, meaning that for any value of R1, R2, or V_{in} we can find the value of V_O. This mathematical relation will remain the same for all the values.

Now, how do we use this transfer function to understand the behavior of the network? Let's say that we want to know how the output voltage, V_O, will change when we change the value of R1.

Let R2 = 10kΩ, V_{in} = 10V, and R1 change from 0 to 20kΩ, in the steps of 2kΩ. From *Equation 5.4*, we know that the output voltage (V_O) is the transfer function (A) times the input voltage (V_{in}). First, we will evaluate the transfer function (*Equation 5.8*) by substituting different values of R1. Mathematically, we can write this as A(R1). The values of A(R1) are calculated as follows:

For R1 = 0Ω:

$$A(R1 = 0\Omega) = \left(\frac{10k}{0+10k}\right) = 1$$

For R1 = 2kΩ:

$$A(R1 = 2k\Omega) = \left(\frac{10k}{2k+10k}\right) = 0.833$$

For R1 = 4kΩ:

$$A(R1 = 4k\Omega) = \left(\frac{10k}{4k+10k}\right) = 0.714$$

*

*

*

For R1 = 20kΩ:

$$A(R1 = 20k\Omega) = \left(\frac{10k}{20k+10k}\right) = 0.333$$

Next, we want to find the output voltage $V_O(R1)$. We will use *Equation 5.4* to find these values. From *Equation 5.4*, we can write:

$$V_O(R1) = A(R1) * V_{in}$$

For R1 = 0Ω:

$$V_O(R1 = 0) = A(0) * 10V = 1 * 10V = 10V$$

For R1 = 2kΩ:

$$V_O(R1 = 2k) = A(2k) * 10V = 0.833 * 10V = 8.33V$$

For R1=4kΩ:

$$V_O(R1 = 4k) = A(4k) * 10V = 0.714 * 10V = 7.14V$$

*
*
*

For R1 = 20kΩ:

$$V_0(R1 = 20k) = A(20k) * 10V = 0.333 * 10V = 3.33V$$

Next, we summarize the calculated values of A(R1) and V_0(R1) in *Table 5.1*. We can plot the tabulated points of *Table 5.1* as shown in *Figure 5.3a*. The plot of *Figure 5.3a* is called A versus R1, or A as a function of R1. From these graphs, we can observe the behavior of the network. We see that :

1) The graph is decreasing in magnitude, when the value of R1 is increased.
2) The maximum gain occurs when R1 = 0 (short).
3) If we make R1 very large, the gain will approach 0.

These types of observations help us understand the behavior of the system better, and help us predict the system's response.

R1(Ω)	**A(R1)**	**V_{in} = 10V V_0(R1) = A(R1) * V_{in}**
0	1	10
2k	0.833	8.33
4k	0.714	7.14
6k	0.625	6.25
8k	0.555	5.55
10k	0.5	5
12k	0.454	4.54
14k	0.417	4.17
16k	0.385	3.85
18k	0.357	3.57
20k	0.333	3.33

Table 5.1

For example, let R1 = 2kΩ, R2 = 10kΩ, and V_{in} = 10V. From the graph of *Figure 5.3a*, we know that for R1 = 2kΩ (and R2 = 10kΩ), the gain of the network is 0.833. We also know from *Equation 5.4* that when we multiply A and V_{in}, we get the output; therefore, 0.833 * 10V gives us 8.33V. So, for a 10V input, this network's response is an output of 8.33V. Now suppose that 8.33V is not desirable, then what do we do? We simply alter the network's behavior (or the transfer function) to obtain the desired value.

Suppose we want the network to produce an output of 6.25V for the input of 10V. Then from the graph of *Figure 5.3a*, we know that by making R1 = 6kW, we get the desired output. Now, let's go one step further and say that we want the output of 6.25V, but for one reason or another, we are not allowed to change the components of the network. What is the next step? We simply add another network to compensate for the difference.

For example, our first network gives us 8.33V as an output, and the second network will accept 8.33V as input and produce an output of 6.25V. These two networks are shown in *Figure 5.3c*.

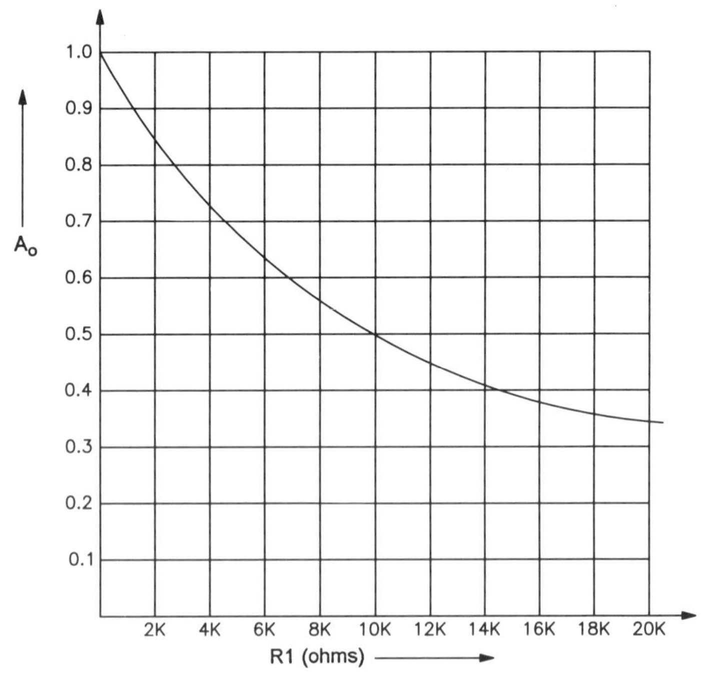

Figure 5.3a

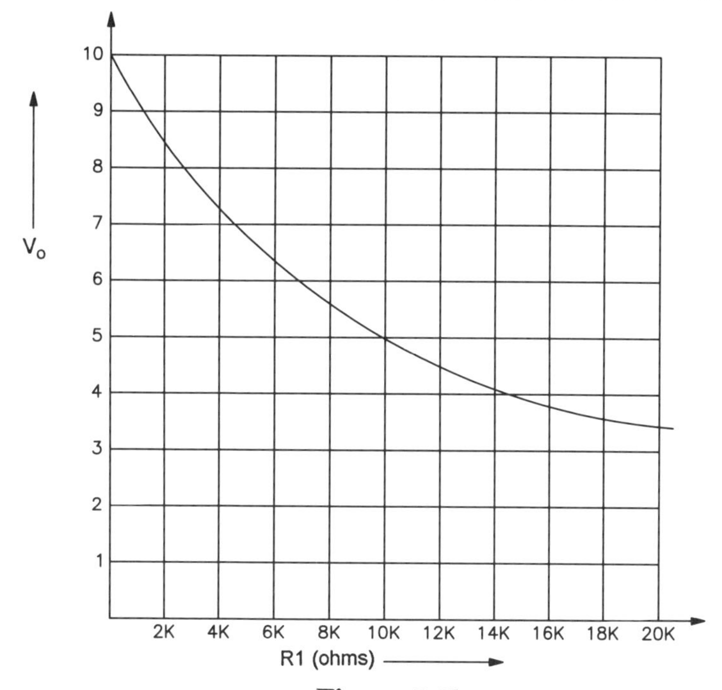

Figure 5.3b

The second network with a transfer function of A_2 is called a compensation network because it compensates for the difference between the desired value and the actual value. Later, we will learn more about compensation networks.

5.2.2 TIME-DOMAIN, FREQUENCY DOMAIN, AND S-DOMAIN TRANSFER FUNCTIONS

In the circuit of *Figure 5.1*, we see that for a given set of resistor values, the gain is constant. For example, for R1 = 2kΩ, R2 = 10kΩ, the gain is 0.833. No matter what type of input voltage we apply (DC or AC), the output is always 0.833 times the input voltage because the resistance does not change. But in networks containing reactive components, that is not the case. These types of networks are both time and frequency dependent, meaning that the gain of the network is different depending upon the type of input that is applied. For these types of networks (or systems), we derive time and frequency dependent transfer functions, so we can learn about the behavior

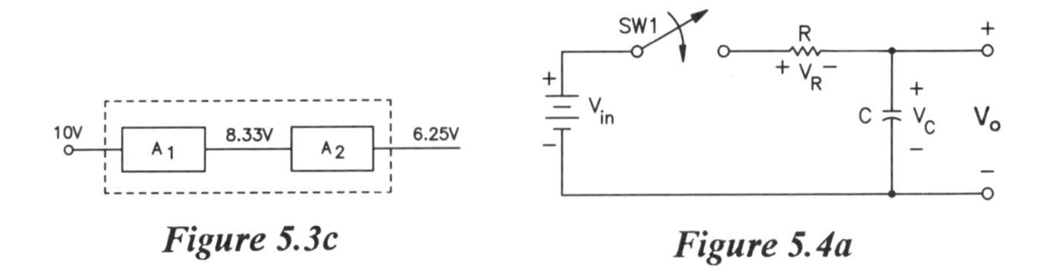

Figure 5.3c

Figure 5.4a

of the network for different types of inputs. These are called time-domain transfer functions and frequency-domain transfer functions. Let's examine these in detail.

Example: R-C Network

Time-Domain Transfer Function

Let's examine the circuit containing a reactive component such as the one shown in *Figure 5.4a.* The time-domain transfer function for this circuit is as follows:

Using Kirchoff's Voltage Law:

$$V_{in} = V_R + V_C$$

(Equation 5.5)

Notice that V_C is parallel to V_0; therefore,

$$V_0 = V_C$$

(Equation 5.6)

Substituting *Equation 5.6* into *Equation 5.5*:

$$V_{in} = V_R + V_0$$

(Equation 5.7)

or

$$V_0 = V_{in} - V_R$$

(Equation 5.8)

In the R-C circuit, the voltage across the resistor is given by:

$$V_R = V_{in} * \left(e^{\left(\frac{-t}{R*c} \right)} \right)$$

(Equation 5.9)

Substituting *Equation 5.9* into *Equation 5.8*:

$$V_0 = V_{in} - \left(V_{in} * e^{\left(\frac{-t}{R*c} \right)} \right) * V_{in}$$

(Equation 5.10)

$$V_0 = \left(1 - e^{\left(\frac{-t}{R*C} \right)} \right) * V_{in}$$

Output Voltage   Transfer Function   Input Voltage

(Equation 5.11)

$$\frac{Vo}{Vin} = \left(1 - e^{\left(\frac{-t}{R*c} \right)} \right)$$

(Equation 5.12a)

$$\text{Let } A = \frac{Vo}{Vin}$$

(Equation 5.12b)

Substitute *Equation 5.12b* into *Equation 5.12a*:

$$A = \left(1 - e^{\left(\frac{-t}{R*c} \right)} \right)$$

(Equation 5.12c)

From the transfer function of *Equation 5.11*, we can see that there are four independent variables: t, R, C, and V_{in}, meaning that any of these variables can change independently, whereas V_0 is a dependent variable which changes only when any of the independent variables change. The transfer function of *Equation 5.11* is called a *time-domain function*. Time-domain functions are studied to understand the system's characteristics as a function of time: For example, how long does it take the output to reach its nominal value, or if the output goes up or down from its nominal value, how long will it take for it to come back to its nominal value, etc.

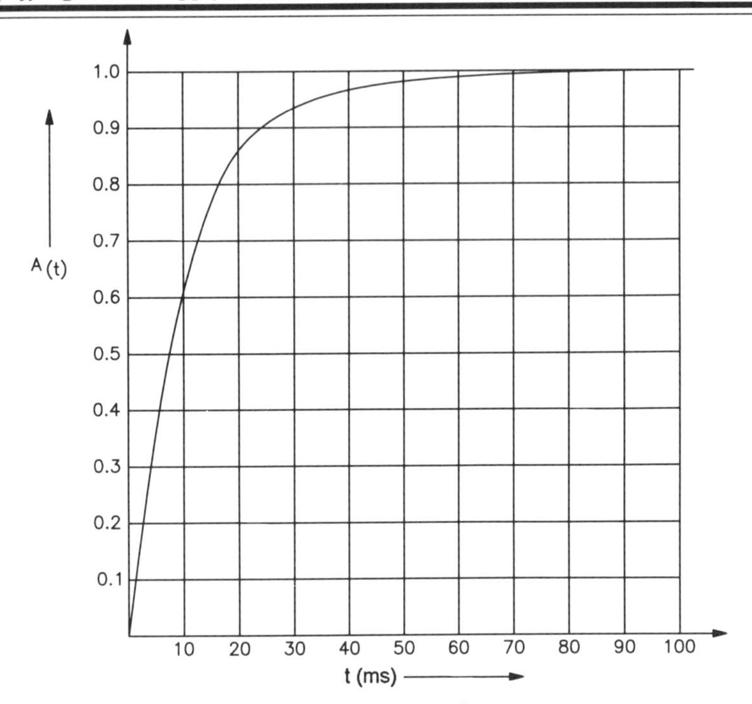

Figure 5.4b

t(s)	A(t)
0	0
2m	0.181
4m	0.33
6m	0.451
8m	0.449
10m	0.632
20m	0.865
30m	0.95
40m	0.982
50m	0.993
60m	0.998
70m	0.999
100m	1

Table 5.2

From the transfer function of *Equation 5.12c*, let's understand the behavior of the circuit (*Figure 5.4a*). We want to see how the gain, A, changes as a function of time, or A(t).

Let R = 10kΩ, C = 1μF, V_{in} = 10V, and close the switch at t = 0:

$$A(t=0) = \left(1 - e^{-\left(\frac{0}{10E3 * 1E-6}\right)}\right) = (1 - e^{0}) = (1 - 1) = 0$$

$$A(t=2ms) = \left(1 - e^{-\left(\frac{2E-3}{10E3 * 1E-6}\right)}\right) = (1 - e^{-0.2}) = (1 - 0.819) = 0.181$$

$$A(t=4ms) = \left(1 - e^{-\left(\frac{4E-3}{10E3 * 1E-6}\right)}\right) = (1 - e^{-0.4}) = (1 - 0.670) = 0.330$$

*

*

*

$$A(t=10ms) = \left(1 - e^{-\left(\frac{10E-3}{10E3 * 1E-6}\right)}\right) = (1 - e^{-1}) = (1 - 0.368) = 0.632$$

$$A(t=20ms) = \left(1 - e^{-\left(\frac{20E-3}{10E3 * 1E-6}\right)}\right) = (1 - e^{-2}) = (1 - 0.135) = 0.865$$

$$A(t=30ms) = \left(1 - e^{-\left(\frac{30E-s}{10E3 * 1E-6}\right)}\right) = (1 - e^{-4}) = (1 - 0.050) = 0.950$$

*

*

*

$$A(t=100ms) = \left(1 - e^{-\left(\frac{100-3}{10E3 * 1E-6}\right)}\right) = (1 - e^{-10}) = (1 - 0) = 1.000$$

The results of the above equations are summarized in *Table 5.2*.

Next, we will plot the tabulated points of *Table 5.2* in *Figure 5.3b*. From the graph, after we close the switch, we observe that:

1) The gain changes with time.
2) It takes 100ms to reach the maximum gain.

To see an output of 10V (for the input voltage = 10v), we must wait for 100ms. We can alter the behavior of this circuit to change the time to 10ms simply by changing the component values in the network.

Frequency-Domain Transfer Function

Normally, time-domain functions are studied to understand the time-related responses of the system. But if we want to know how the system behaves under different frequencies, then we study the *frequency-domain transfer functions*: For example, how does the system behave when we apply different frequencies, etc. Next we will derive a frequency-domain transfer function of the circuit of *Figure 5.5.*

In the previous section, the input voltage, V_{in}, was a constant value of DC. In this section, we will make V_{in} a sine wave with a constant amplitude but varying frequency.

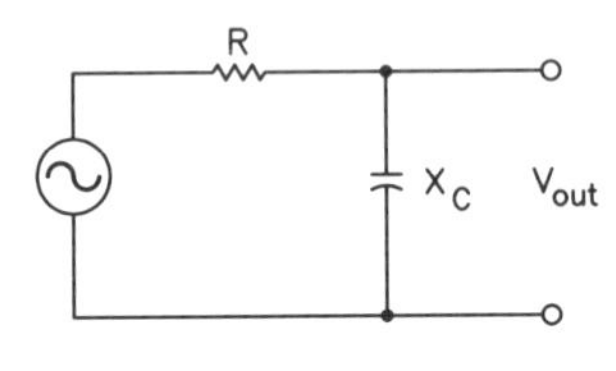

Figure 5.5

A Brief Review

Recall that the reactance of C is given by X_C:

$$X_C = |X_C| \angle -90, \text{ where } |X_C| \text{ is the amplitude}$$
$$\text{and } \angle -90 \text{ is the phase angle.}$$

(Equation 5.13a)

The angle comes from the fact that a reactive component creates a 90° phase shift between the voltage and current. The (-) sign comes because the voltage is lagging the current. (In the inductive circuit $X_L = |X_L| \angle +90$, the (+) sign comes because the voltage is leading the current. Recall the phrase, "ELI is the ICE man.") The expression in *Equation 5.13a* is given in polar form. Another way to represent it is in rectangular form:

$$X_C = -jX_C, \text{ where } -j \text{ is } -90° \text{ degrees.}$$

(-j is also called an imaginary term.)

(Equation 5.13b)

$$\text{Also, } -j = 1/j$$

(Equation 5.13c)

Therefore, $X_C = 1/(j\omega C)$
(Equation 5.13d)

Recall that $\omega = 2 * \pi * f$, where f is the frequency in Hz
and ω is in radians/seconds.
(Equation 5.14)

Redrawing the circuit of *Figure 5.4a* to the one in *Figure 5.5:* to find V_O, we need to find the voltage across X_C. We can treat this circuit as a voltage divider. For example:

$$V_0 = \frac{Xc}{R+Xc} * V_{in}$$
(Equation 5.15)

$$\frac{Vo}{Vin} = \frac{Xc}{R+Xc}$$
(Equation 5.16)

The transfer function of *Equation 5.16* contains a real (resistor) term and an imaginary (reactance) term; therefore, the transfer function must be calculated both in gain and phase. Let's look at the reasoning behind it.

In the resistor divider network, the voltage divider ratio remains constant because the input voltage source is constant in frequency, and the resistance does not depend on frequency; whereas in the R-C network, the capacitive reactance changes as a function of frequency. Therefore, the gain changes. Also, because of the capacitive reactance, there is a phase shift involved. Recall that in a purely capacitive network, the current leads the voltage. So, in our R-C network, the voltage across the capacitor is the product of the current through it and the reactance, which is frequency dependent. Substituting *Equation 5.13d* into *5.16*:

$$\frac{Vo}{Vin} = \frac{\left(\frac{1}{j*\omega*C}\right)}{R+\left(\frac{1}{j*\omega*C}\right)}$$
(Equation 5.17a)

$$\frac{Vo}{Vin} = \frac{\left(\frac{1}{j*\omega*C}\right)}{\left(\frac{R*(j*\omega*C)+1}{j*\omega*C}\right)}$$
(Equation 5.17b)

Canceling out the $j * \omega * C$ term from both the numerator and the denominator, we can simplify *Equation 5.17b* as follows:

$$\frac{Vo}{Vin} = \frac{1}{1+(j*\omega*R*C)}$$

(Equation 5.17c)

$$\text{Let } fo = \frac{1}{2*\pi*R*C}$$

(Equation 5.17d)

f_0 is called the corner frequency, which we will look at shortly. Substituting *Equation 5.17d* and *Equation 5.14* into *Equation 5.17c*:

$$\frac{Vo}{Vin} = \frac{1}{1+\frac{j*2*\pi*f}{2*\pi*fo}}$$

(Equation 5.17e)

Canceling out the $2 * \pi$ terms from *Equation 5.17e*:

$$\frac{Vo}{Vin} = \frac{1}{1+j*\frac{f}{fo}}$$

(Equation 5.17f)

$$\text{Let } A = \frac{Vo}{Vin}$$

(Equation 5.17g)

Substituting *Equation 5.17g* into *Equation 5.17f*:

$$A = \frac{1}{1+j*\frac{f}{fo}}$$

(Equation 5.17h)

Since the transfer function of *Equation 5.17h* contains a real and an imaginary term, the magnitude and the phase can be calculated by the following relations. To find the magnitude of the gain in *Equation 5.17h*, we use the following equation:

$$|A| = \frac{1}{1+\left(\frac{f}{fo}\right)^2}$$

(Equation 5.17i)

To find the phase angle of *Equation 5.17h*, we use the following equation:

$$\angle A = \text{-arctangent}\left(\frac{f}{fo}\right) \text{ or } \angle A = -\tan^{-1}\left(\frac{f}{fo}\right)$$

(Equation 5.17j)

By substituting different values of *f* in *Equation 5.17i* and *Equation 5.17j*, we can generate gain and phase characteristics of the network. The gain expression is commonly written in logarithmic form as shown below.

Let's convert *Equation 5.17i* into the log form. Multiply both sides of the equations by 20 * $\log_{10}$. We will write $\log_{10}$ simply as *log*.

$$20\log|A| = 20\log\left(\frac{1}{\sqrt{1+\left(\frac{f}{fo}\right)^2}}\right)$$

(Equation 5.17k)

At this point, let's review some identities of *log* functions:

$$\log(A * B) = \log A + \log B$$

$$\log(A/B) = \log A - \log B$$

$$\log\sqrt{A} = \log(A)^{\frac{1}{2}} = \left(\frac{1}{2}\right) * \log(A)$$

Substituting the above identities into *Equation 5.17k*:

$$20\log|A| = 20\log(1) - 20\log\sqrt{1+\left(\frac{f}{fo}\right)^2}$$

(Equation 5.17l)

$$20\log|A| = 20\log(1) - \left(\frac{1}{2}\right) * 20\log\left(1+\left(\frac{f}{fo}\right)^2\right)$$

(Equation 5.17m)

The left side of the equation is expressed as A_{dB}, where dB stands for decibel. Recall that log(1) = 0; therefore, *Equation 5.17m* can be rewritten as:

$$A_{dB} = -10\log\left(1+\left(\frac{f}{fo}\right)^2\right)$$

(Equation 5.17n)

Let R = 1kΩ and C = 0.1μF. First, we calculate the corner frequency, f_o. From *Equation 5.17a*:

$$f_0 = \frac{1}{2*\pi*R*C}$$

(Equation 5.17o)

$$f_0 = \frac{1}{2*\pi*(1e+3)*(1e-6)} = 159Hz$$

(Equation 5.17p)

Next, we want to find the magnitude and phase of the system as a function of the frequency. To find the magnitude of the gain A(f), we use *Equation 5.17n*:

$$A_{dB} = -10\log\left(1+\left(\frac{f}{f_0}\right)^2\right)$$

(Equation 5.17q)

Substitute *Equation 5.17q* into *Equation 5.17p*:

$$A_{dB} = -10\log\left(1+\left(\frac{f}{159}\right)^2\right)$$

(Equation 5.17r)

For $f = 0$:

$$A_{dB}(f=0) = -10\log\left(1+\left(\frac{0}{159}\right)^2\right) = -10\log(1+0) = 0_{dB}$$

(Equation 5.17s)

$$A_{dB}(f{=}100Hz){=}{-}10\log\left(1+\left(\frac{100}{159}\right)^2\right){=}{-}10\log(1+0.396){=}{-}1.45_{dB}$$

$$A_{dB}(f{=}159Hz){=}{-}10\log\left(1+\left(\frac{159}{159}\right)^2\right){=}{-}10\log(1+1){=}{-}3_{dB}$$

$$A_{dB}(f{=}200Hz){=}{-}10\log\left(1+\left(\frac{200}{159}\right)^2\right){=}{-}10\log(1+1.58){=}{-}4.12_{dB}$$

$$A_{dB}(f{=}400Hz){=}{-}10\log\left(1+\left(\frac{400}{149}\right)^2\right){=}{-}10\log(1+2.52){=}{-}8.65_{dB}$$

$$A_{dB}(f{=}600Hz){=}{-}10\log\left(1+\left(\tfrac{600}{159}\right)^2\right){=}{-}10\log(1+14.24){=}{-}11.83_{dB}$$
$$A_{dB}(f{=}800Hz){=}{-}10\log\left(1+\left(\tfrac{800}{159}\right)^2\right){=}{-}10\log(1+25.32){=}{-}14.20_{dB}$$
$$A_{dB}(f{=}100Hz){=}{-}10\log\left(1+\left(\tfrac{1000}{159}\right)^2\right){=}{-}10\log(1+39.56){=}{-}16.08_{dB}$$

*

*

*

To find the phase angle, we use *Equation 5.17j*:

$$\angle A(f=0) = -\tan^{-1}\left(\tfrac{0}{159}\right) = -\tan^{-1}(0) = 0^o$$
$$\angle A(f=100Hz) = -\tan^{-1}\left(\tfrac{100}{159}\right) = -\tan^{-1}(0.629) = -32.17^o$$
$$\angle A(f=159Hz) = -\tan^{-1}\left(\tfrac{159}{159}\right) = -\tan^{-1}(1) = -45.0^o$$
$$\angle A(f=200Hz) = -\tan^{-1}\left(\tfrac{200}{159}\right) = -\tan^{-1}(1.26) = -51.51^o$$
$$\angle A(f=400Hz) = -\tan^{-1}\left(\tfrac{400}{159}\right) = -\tan^{-1}(2.52) = -68.32^o$$
$$\angle A(f=600Hz) = -\tan^{-1}\left(\tfrac{600}{159}\right) = -\tan^{-1}(3.77) = -75.16^o$$
$$\angle A(f=800Hz) = -\tan^{-1}\left(\tfrac{800}{159}\right) = -\tan^{-1}(5.03) = -78.76^o$$
$$\angle A(f=1000Hz) = -\tan^{-1}\left(\tfrac{1000}{159}\right) = -\tan^{-1}(6.29) = -80.97^o$$

*

*

*

Frequency (Hz)	Magnitude of Gain A(f)dB	Phase Angle of A (degress)
0	0	0
100	-1.45	-32.17
159	-3	-45
200	-4.12	-51.51
400	-8.65	-68.32
600	-11.83	-75.16
800	-14.2	-78.76
1,000	-16.08	-80.96
2,000	-22.02	-85.45
4,000	-28.02	-87.72
6,000	-31.54	-88.48

Table 5.3

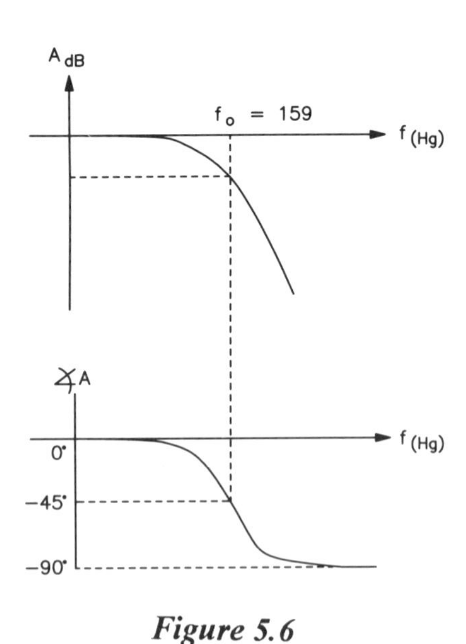

Figure 5.6

Next, we summarize the calculated values in *Table 5.3*. The results of the table can be plotted in *Figure 5.6*.

The graphs of *Figure 5.6* are called *gain vs. frequency* and *phase vs. frequency* plots. The horizontal axis and the vertical axis are normally shown in log scales. These graphs are also known as *Bode plots,* named after developer Hendrik W. Bode. He was a mathematician and an electrical engineer.

In *Figure 5.6*, we see that the gain is decreasing at a certain rate, which is called the roll-off rate. In our R-C network, it is decreasing at -20dB per decade. The phase is decreasing at -45° per decade; however, the phase angle goes from 0 to -90°. At the corner frequency, it is -45°. The circuit described in *Figure 5.6* is also called a single-pole circuit.

So, for the R-C network, we derive a time-domain function and a frequency-domain function.

s-Domain Transfer Function

At this point, we will briefly look at the s-domain. The "*s*" comes from the technique called LaPlace transform. In this technique, the complex term jw is replaced by *s*, or $s = j\omega$ for a constant amplitude sine wave. It should be emphasized that there is a great deal involved in making this substitution, but for the purpose of explaining the transfer functions in s-domain, we will replace $j\omega$ with *s* (while keeping in mind that *s* is also a complex term). The s-domain transfer function is also called the *LaPlace-domain transfer function.*

For example, *Equation 5.13d* can be rewritten in s-domain as $X_C = 1/sC$.

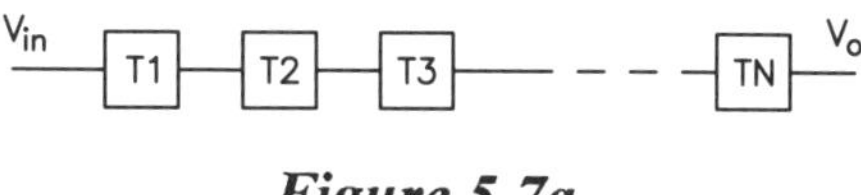

Figure 5.7a

Basic Operations of Transfer Functions

Let's look at some basic operations of transfer functions.

When two or more transfer functions, T_1, T_2, T_3, etc., are as shown in *Figure 5.7a*, then:

$$V_O = (T_1 * T_2 * T_3 * \ldots\ldots * T_N) * V_{in}$$

(Equation 5.18)

When the transfer functions are as shown in *Figure 5.7b*, then:

$$V_O = (T_1 + T_2 + T_3 \ldots\ldots + T_N) * V_{in}$$

(Equation 5.19)

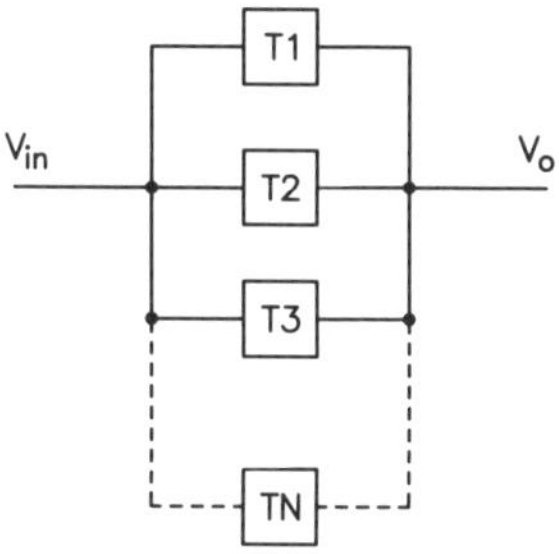

Figure 5.7b

If T_1 was the transfer function for the R-C circuit, called T_{1R-C} (see *Equation 5.17h*), and T_2 was the transfer function of another R-C circuit, T_{2R-C}, then:

$$V_O = (T_1) * (T_2) * V_{in}$$
(Equation 5.20a)

$$V_O = (T_{1R-C}) * (T_{2R-C}) * V_{in}$$
(Equation 5.20b)

A combined transfer function is:

$$V_O = (T_{1R-C} * T_{2R-C})$$
(Equation 5.21)

5.3 CLOSED-LOOP CONTROL SYSTEM

Next we will look at a transfer function of a basic closed-loop control system or feedback control system (*Figure 5.8a*).

$$Y = e * G$$
(Equation 5.22a)

$$c = H * Y$$
(Equation 5.22b)

$$e = X - (c)$$
(Equation 5.23)

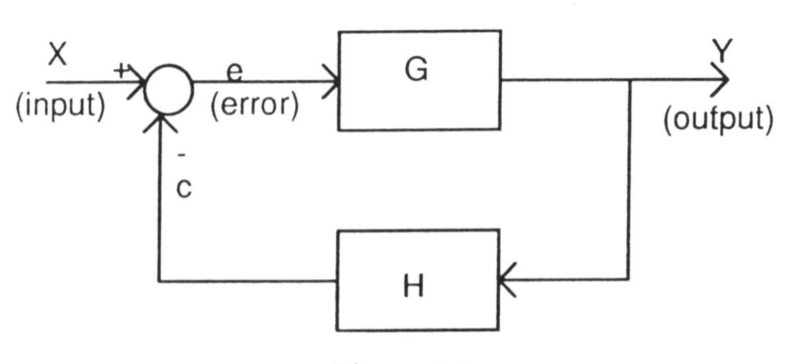

Figure 5.8a

Substituting *Equation 5.22b* and *Equation 5.23* into *Equation 5.22a*:

$$Y = (X - H * Y) * G$$

(Equation 5.24)

Rearranging *Equation 5.24*:

$$Y = X * G - H * Y * G$$

(Equation 5.25)

Combining the Y terms in *Equation 5.25*:

$$Y + H * Y * G = X * G$$

(Equation 5.26)

$$Y(1 + G * H) = G * X$$

(Equation 5.27)

$$Y = \left(\frac{G}{1+G*H}\right) * X$$

Output Transfer Function Input

(Equation 5.28a)

$$\text{Let } A = \frac{Y}{X}$$

(Equation 5.28b)

Substitute *Equation 5.28b* into *Equation 5.28a*:

$$A = \left(\frac{G}{1+G*H}\right)$$

(Equation 5.29)

Equation 5.29 is called a *closed-loop transfer function*, and the GH in the denominator is called an *open-loop transfer function.* It can be written in time-domain, frequency-domain, or S-domain.

For example, in s-domain, the transfer function of *Equation 5.29* will be written as follows:

$$A(s) = \frac{G(s)}{1+G(s)*H(s)}$$

(Equation 5.30)

Let's say that the G-block contains the R-C circuit of *Figure 5.5* (also see *Equation 5.17c*), and the H-block contains a constant gain of K; then we can rewrite the above transfer function as:

$$G = \frac{1}{1+s*R*C}$$

(Equation 5.31a)

$$H = K$$

(Equation 5.31b)

$$Y = \frac{G}{1+GH} * X$$

(Equation 5.31c)

$$Y = \frac{\left(\frac{1}{1+s*R*C}\right)}{\left(1+\left(\frac{1}{1+s*R*C}*K\right)\right)} * X$$

(Equation 5.31d)

Let $A = \frac{Y}{X}$; therefore, the above equation becomes:

$$A = \frac{\left(\frac{1}{1+s*R*C}\right)}{\left(1+\left(\frac{1}{1+s*R*C}*K\right)\right)}$$

(Equation 5.32)

The above expression is also called a closed-loop transfer function of the control system, and the general response is plotted in *Figure 5.8b* using the method shown in the previous examples. The curves of *Figure 5.8b* can be shifted up/down or left/right simply by changing the values of R and/or C; therefore, the location of the corner frequency can be changed. The curve also shows the slope at which the gain is decreasing. In many systems it is desirable to have a steeper slope. This can be achieved by adding more stages containing poles. Also, the phase angle becomes more negative with a higher number of poles.

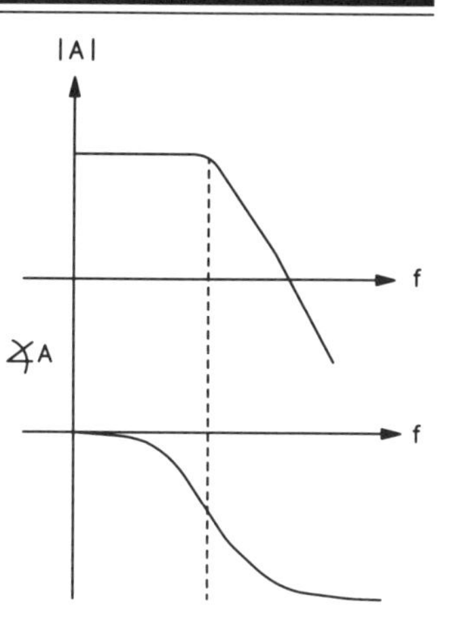

Figure 5.8b

By analyzing the Bode plots of the closed-loop system, we can learn about the behavior of the system and answer questions like, "Will the output oscillate or will the output be stable?" There are various techniques used for graphically representing the transfer functions. In our examples, we have only looked at the Bode plot technique. Other techniques are the *Nyquist plot* and the *Root-locus plot.* (Readers may refer to other books on these subjects.)

Let's use a Bode plot to understand if the system will be stable. To do this we will perform a simple test. Examining a Bode plot, when the gain of the system crosses the 0_{dB} point, if at that point the phase is -180°, then the system will be unstable. If this is not the case, then it is a stable system. We see from the closed-loop response (as shown in *Figure 5.8c*) of our basic control system that when the gain curve crosses the 0_{dB} line, the phase angle is -140°. Since it is not -180°, we can say that the feedback system is stable. At this point the phase difference between -180° and -140° is called the *phase margin.* Another test is that when the phase reaches -180°, the gain must be below 0_{dB}. This is called the *gain margin.*

Now, suppose that at a 0 crossover, the phase angle is -180°. If this is the case, then in our H-block we can insert a compensation network. This compensation network will add an extra transfer function in the loop and change the overall closed-loop transfer function. With the compensation network, we can alter the characteristics of the closed-loop system. This is done by simply changing the components of the network. In section 5.4.2 we will discuss various types of compensation networks.

We have examined how to use the frequency-domain transfer function to predict the stability of the system. There are many other characteristics that can be predicted by using various techniques. For example, to find out how long it will take for the feedback loop to bring the output down to its original value after the output voltage increases, we can use the time-domain function. In effect, these mathematical expressions and graphical

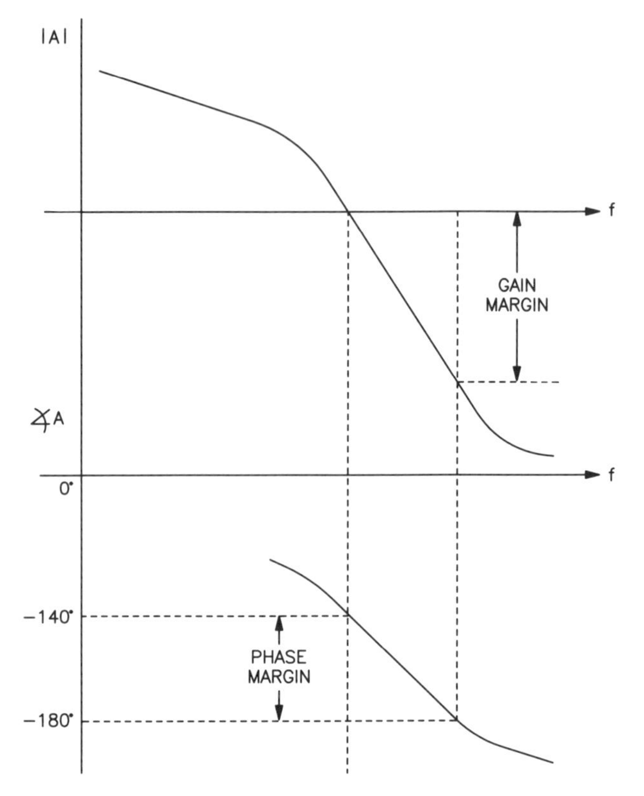

Figure 5.8c

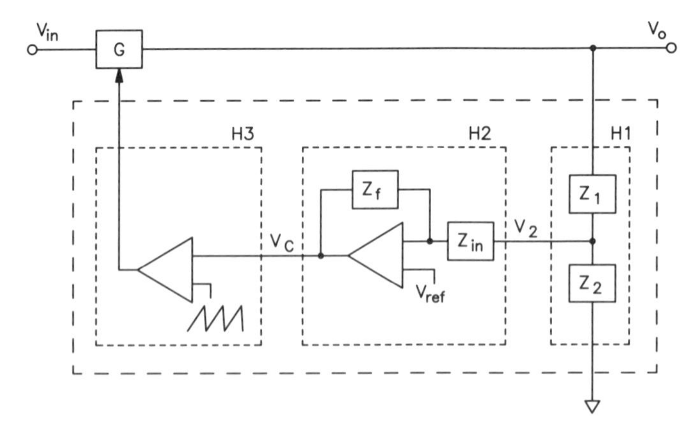

Figure 5.9a

representations are used as tools. These tools may seem tedious and long, but most of these functions can be performed on computers by using various software packages.

Now we will see how the components in the H-block (control section) are implemented. The details of the control section are shown in *Figure 5.9a.*

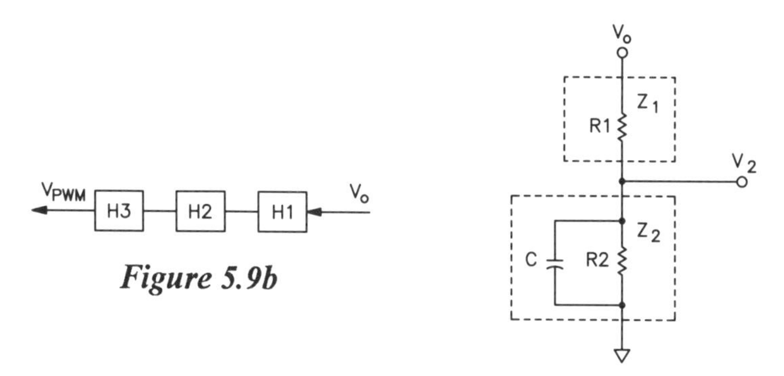

Figure 5.9b

Figure 5.9c

5.4 FEEDBACK CIRCUIT

5.4.1 VOLTAGE DIVIDER NETWORK

The output voltage is connected to the voltage divider network, and the output of the H-1 block is connected as an input of the error amplifier and compensation network. We will call this the H-2 block. The output of the H-2 block is then connected as an input to a comparator which generates PWM signals. We will call this the H-3 block. The simplified block diagram is shown in *Figure 5.9b.*

The voltage divider transfer function containing only resistors was derived in *Equation 5.1.* We can expand on it and say that an equation for the circuit of H-2 block in *Figure 5.9a* can be written as:

$$V_2/V_O = Z_2/(Z_1+Z_2)$$

(Equation 5.33)

The input to the H-1 block is V_O and the output is V_2; therefore, the output to input ratio is V_2/V_O.

Z_2 in *Figure 5.9a* can be any combination of resistors and capacitors. Let's say that the Z_1 block only contains a resistor and Z_2 contains a resistor in parallel with a capacitor (see *Figure 5.9c*). The voltage divider ratio can then be written as:

$$\frac{V2}{Vo} = \frac{R1}{R1+Z2}$$

(Equation 5.34)

$$Z_2 = R_2 / X_C; X_C = \frac{1}{s*C}$$

(Equation 5.35)

Substituting *Equation 5.35* into *Equation 5.34*:

$$Z_2 = \frac{(R1)(Xc)}{R1+Xc}$$

(Equation 5.36)

$$\text{Where } X_c = \frac{1}{s*C}$$

(Equation 5.37)

$$Z_2 = \frac{(R1)*\left(\frac{1}{s*C}\right)}{R1+\left(\frac{1}{s*C}\right)}$$

(Equation 5.38)

Equation 5.38 can be reduced to:

$$Z_2 = \frac{R1}{1+(s*C*R1)}$$

(Equation 5.39)

Equation 5.39 looks like a simple algebraic equation, but it contains a complex term (j).

5.4.2 COMPENSATION NETWORK

We will look at some examples with an op-amp. The H-2 block contains the compensation network, and the compensation network normally contains an op-amp circuit.

For the circuit of *Figure 5.10a*, the transfer function can be written as:

$$\frac{V2}{Vc} = -\frac{Rf}{Rin}$$

(Equation 5.40)

This is an inverting amplifier. V_C is opposite in polarity from V_2. For that reason, the (-) sign is present in the transfer function.

If we replace R_f and R_{in} with Z_f and Z_{in}, respectively, (where Z is the impedance), we will have the same relation as shown in *Equation 5.40* (see also *Figure 5.10b*). Rewriting *Equation 5.40*:

$$\frac{V_2}{V_c} = -\frac{Zf}{Zin}$$

(Equation 5.41)

Z_f and Z_{in} can be made up of any combination of R's and C's. Let's write a transfer function of the circuit in *Figure 5.10c*. Recall that:

$$X_C = \frac{1}{sC}$$

(Equation 5.42)

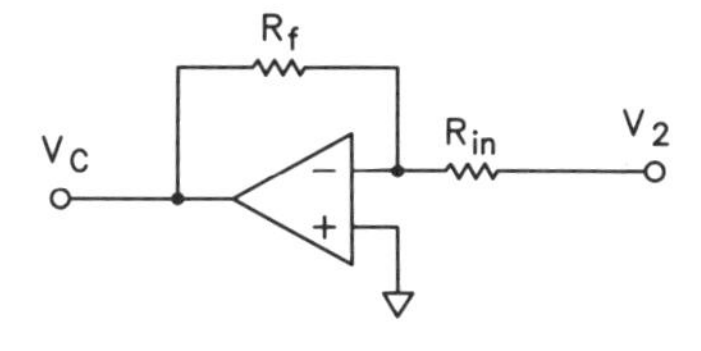

Figure 5.10a

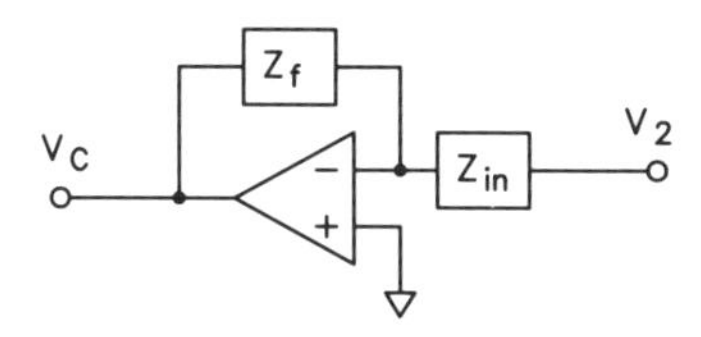

Figure 5.10b

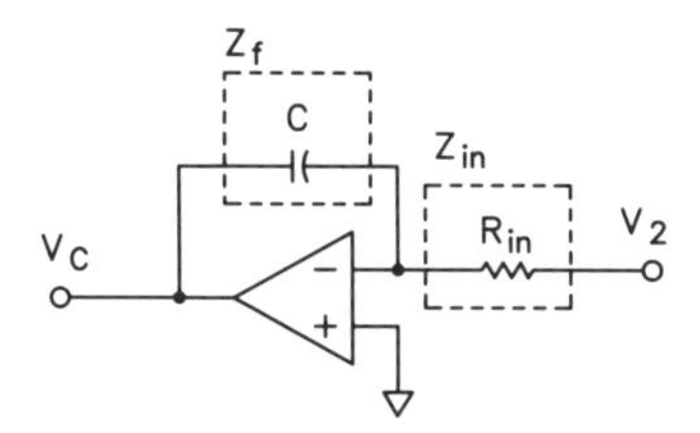

Figure 5.10c

Therefore:

$$\frac{V_c}{V_2} = -\frac{Zf}{Zin} = -\frac{\left(\frac{1}{s*C}\right)}{Rin} = -\frac{1}{s*Rin*C}$$

(Equation 5.43)

Let's derive a transfer function for the circuit in *Figure 5.11a*:

$$\frac{V_c}{V_2} = -\frac{Zf}{Zin}$$

(Equation 5.44a)

$$Z_f = R_f + X_{Cf} = R_f + \frac{1}{s*Cf}$$

(Equation 5.44b)

$$Z_{in} = R_{in}$$

(Equation 5.44c)

Therefore, substituting *Equation 5.44b* and *Equation 5.44c* into *Equation 5.44a*:

$$\frac{V_c}{V_2} = -\frac{\left(Rf+\frac{1}{s*Cf}\right)}{Rin} = -\frac{(Rf)(s*Cf)+1}{Rin*s*Cf} = -\frac{(1+s*Rf*Cf)}{Rin*s*Cf}$$

(Equation 5.44d)

By inserting R_f in the feedback path of the circuit, we were able to change the characteristics of the overall loop. By combining the transfer functions of the G-block, H-1 block, H-2 block, and H-3 block, we can derive a transfer function of the total closed-loop control system, and by adjusting the poles and zeros of the transfer function, we can control the behavior of our system.

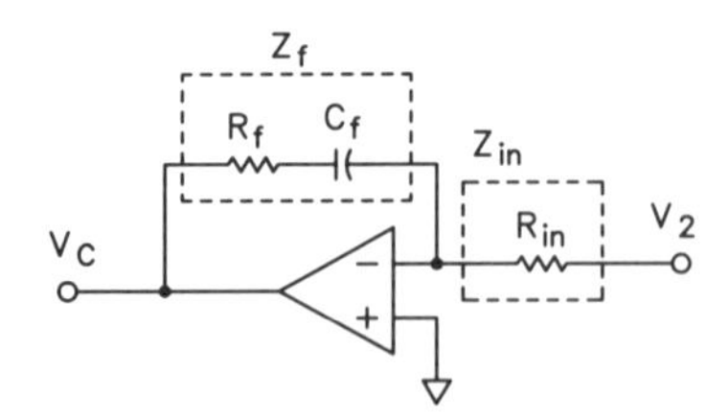

Figure 5.11a

***Example*:**

(See *Figures 5.11b*, *5.11c*, *5.11d*, and *5.11e*).

For the circuit of *Figure 5.11b*:

$$\text{Let } Z_f = R_f // X_{cf}$$

$$\text{Therefore, } Z_f = \frac{(Rf)(Xcf)}{Rf+Xcf}$$

$$\text{Let } X_{cf} = \frac{1}{SCf}$$

$$Z_f = \frac{(Rf)(\frac{1}{SCf})}{Rf+(\frac{1}{SCf})} = \frac{\left(\frac{Rf}{SCf}\right)}{\left(\frac{SR+Cf+1}{CSf}\right)}$$

$$Zf = \left(\frac{Rf}{SRfCf+1}\right)$$

$$\frac{Vc}{V2} = -\frac{Zf}{Zin} = -\frac{Zf}{Rin} = -\frac{\left(\frac{Rf}{SRfCf+1}\right)}{Rin}$$

$$\frac{Vc}{V2} = -\left(\frac{Rf}{Rin}\right)\left(\frac{1}{SRfCf+1}\right)$$

By visual inspection, it can be seen from *Figure 5.11c* that the pole frequency occurs at $f_P = \frac{1}{2\pi RfCf}$.

For the circuit of *Figure 5.11d*:

$$\text{Let } Z_{f1} = R_1 + X_{cf}$$

(1)

$$\text{Let } X_{cf} = \frac{1}{SCf}$$

(2)

$$Z_{f1} = R_1 + \frac{1}{SCf}$$

(3)

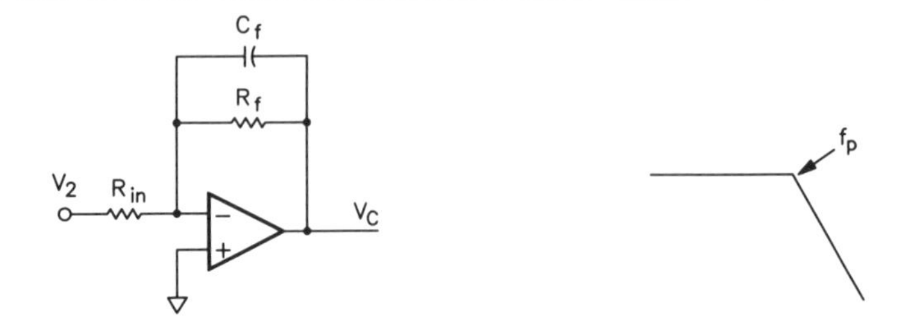

Figure 5.11b ***Figure 5.11c***

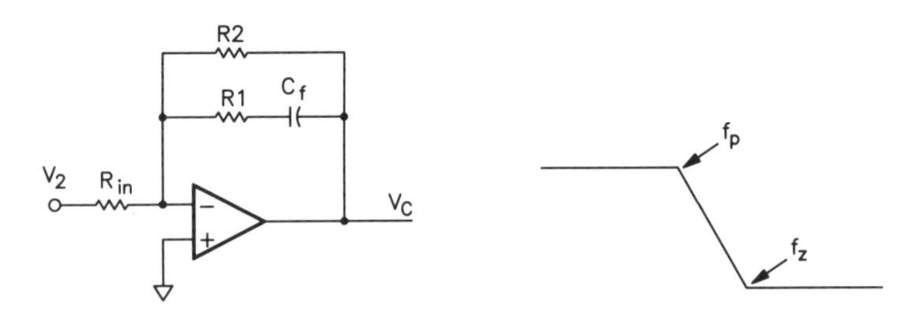

Figure 5.11d ***Figure 5.11e***

$$Z_{f1} = \left(\frac{SR1Cf+1}{SCf}\right)$$
(4)

$$\text{Let } Z_{f2} = R_2 // Z_{f1}$$
(5)

$$\text{Therefore, } Z_{f2} = \frac{(R2)(Zf1)}{R2+Zf1}$$
(6)

Substitute ***(4)*** into ***(6)***:

$$Z_{f2} = \frac{R2\left(\frac{SR1Cf+1}{SCf}\right)}{R2+\left(\frac{SR1Cf+1}{SCf}\right)} = \frac{R2\left(\frac{SR1Cf+1}{SCf}\right)}{\left(\frac{SR2Cf+SR1Cf+1}{SCf}\right)}$$
(7, 8)

$$Z_{f2} = \left(\frac{R2(SR1Cf+1)}{S(R1+R2)Cf+1}\right)$$
(9)

$$\text{To find } \frac{Vc}{V2} = -\frac{Zf2}{Rin} = -\frac{\left(\frac{R2(SR1Cf+1)}{S(R1+R2)Cf+1}\right)}{Rin} = \left(-\frac{R2}{Rin}\right)\left(\frac{SR1Cf+1}{S(R1+R2)Cf+1}\right)$$

(10, 11)

From ***(11)***, we can see that the zero-frequency occurs at $f_Z = \frac{1}{2\pi R1Cf}$, and the pole-frequency at $f_P = \frac{1}{2\pi(R1+R2)Cf}$. If we assume that $R_2 > 0\Omega$, then $R_1 + R_2$ is always higher than R_1. Therefore, f_P is always lower than f_Z. (See *Figure 5.11e*.)

5.5 PULSE-WIDTH MODULATION (PWM)

The majority of today's power supplies are of the *pulse-width modulated* (PWM) type. In this section, we will examine the basic concepts of this technique. We have looked at ON states and OFF states of regulator circuits and converter circuits. Now let's examine these states in detail. How are these states determined? The ON and OFF signals to the switching devices (Qs) are generated by a technique called *pulse-width modulation*, or PWM.

The pulse width is the duration of the ON time (t_{on}) of a switching cycle waveform, as shown in *Figure 5.12a*.

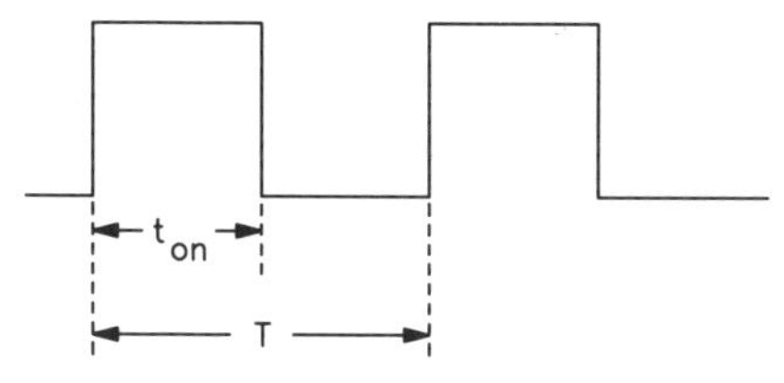

Figure 5.12a

Example: General

Duty cycle, D, is the ratio of t_{on} to T, or:

$$D = \frac{ton}{T}$$

(Equation 5.45)

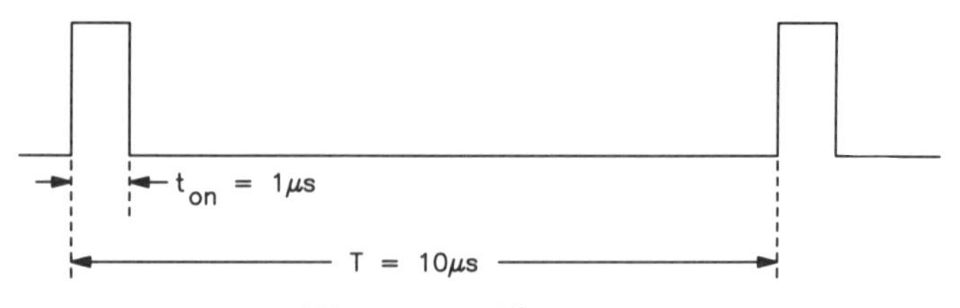

Figure 5.12b

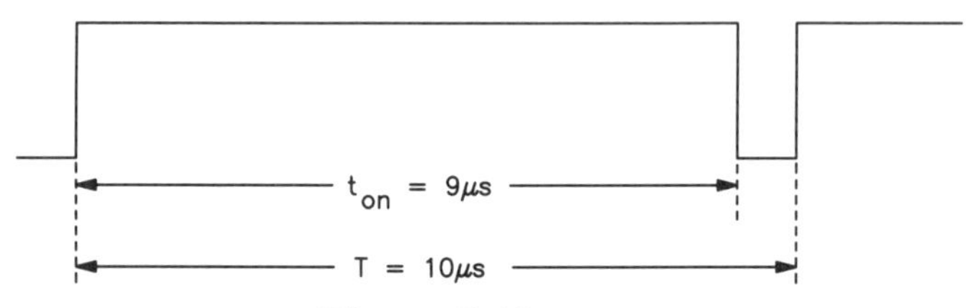

Figure 5.12c

D is usually defined in percentage. For example, if t_{on} = 1μs long and T = 10μs, then the ratio of t_{on} to T is:

$$\frac{ton}{T} = \frac{1\mu s}{10\mu s} = 0.1$$

(Equation 5.46)

This ratio multiplied by 100 gives us the percentage; therefore, 0.1 * 100 = 10%, so the duty cycle is 10% (see *Figure 5.12b*).

Now, suppose t_{on} changes to 9μs and T = 10μs; then the ratio would be:

$$\frac{ton}{T} = \frac{9\mu s}{10\mu s} = 0.9$$

(Equation 5.47)

The duty cycle is 0.9 * 100 = 90% (see *Figure 5.12b*). Ideally, the duty cycle can change:

From 0% ($\frac{0}{10\mu s}$ * 100 = 0%) to 100% ($\frac{10\mu s}{10\mu s}$ * 100 = 100%).

Since the period remains constant and only the pulse width (or t_{on}) changes, the pulse-width is said to be modulating; therefore, this technique is called *fixed-frequency pulse-width modulation*, or simply PWM.

Let's look at the equations of the Buck and Boost regulators that we discussed in Chapter 3:

Example: Buck Regulator

$$V_0 = V_{in} * D$$

(Equation 5.48)

Recall that $D = \frac{ton}{T}$ for 0% duty cycle D = 0

Therefore, $V_O = V_{in} * 0$, or $V * 0 = 0$. So, for the 0% duty cycle, the output voltage is 0; for 100% duty cycle, the duty cycle (D) = 1. Therefore, $V_O = V_{in} * 1$, or $V_O = V_{in}$. So, for the 100% duty cycle, the output voltage is equal to the input voltage.

Example: Boost Regulator

$$V_0 = \frac{1}{1-D} * V_{in}$$

(Equation 5.49)

For the 0% duty cycle, the ratio = $\frac{1}{1-0} = 1$.

Therefore, $V_O = 1 * V_{in}$, or the output voltage is equal to the input voltage.

For 100% duty cycle, the ratio = $\frac{1}{1-1} = \frac{1}{0} = \infty$.

Notice that when the duty cycle is 100% in the Boost regulator, the mathematical expression (*Equation 5.49*) suggests that the output will become infinity or extremely high. But in the practical system, the duty cycle of 1, means that the Q in *Figure 3.7* (Chapter 3) remains in the ON state and the current though Q will become infinity because after a certain time, the L of *Figure 3.7* will charge up and act as a short. The current through Q will exceed the recommended current limit of the device and it will fail. So, in the practical circuit, D = 1 may cause some problems in the circuit.

In previous examples, we saw that the output can be controlled by controlling the ratio of the duty cycle, D, or simply by controlling t_{on} or the pulse width. We also saw, mathematically, how the pulse width controls the output voltage. Now, how do we achieve this pulse-width control using electronic components? *Figure 5.13* shows the basic circuit that achieves PWM. A sawtooth ramp is connected to the (+) terminal of a comparator, and DC voltage is connected to the (-) terminal. In *Figure 5.13*, both of these signals are superimposed. At t_O, the voltage at the (+) terminal is lower than the voltage at the (-) terminal; therefore, the output of the comparator will be low and it will remain low until t_1. Just after t_1, the voltage at the (+) terminal becomes higher than the (-) terminal; therefore, the output of the comparator will be high, and it will remain high until t_2. Just after t_2, the voltage at the (+) terminal drops below the voltage at the (-) terminal, and the output of the comparator goes to low. It will remain low until t_3; however, at T, a period for cycle is ended; therefore, one PWM cycle is completed at T and another begins. To calculate the duty cycle for this particular cycle, we can simply divide the value of the ON time by the value of the period T.

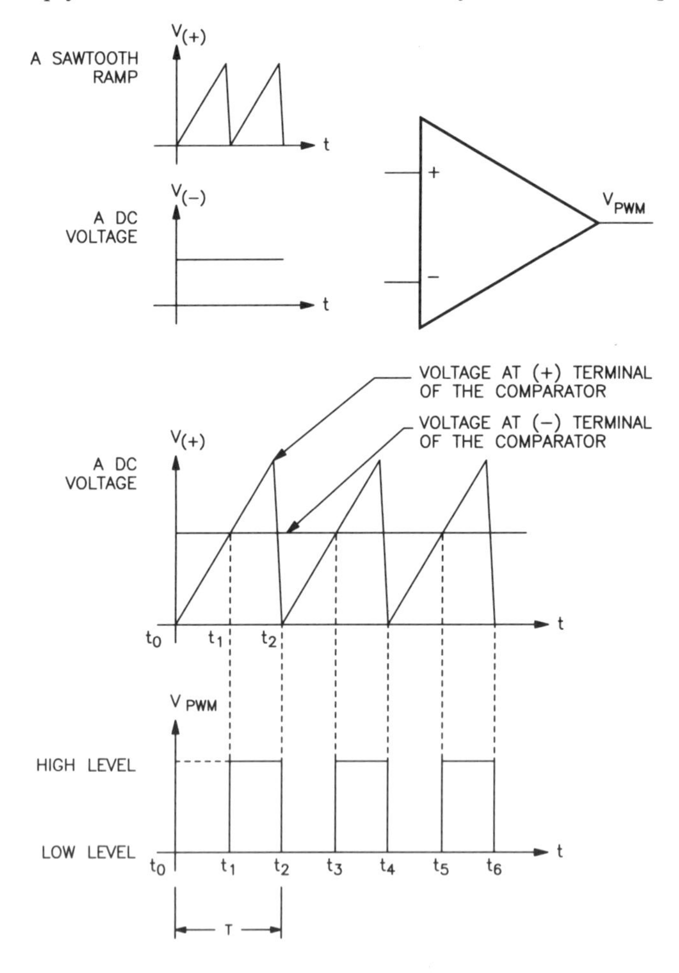

Figure 5.13

For example, the ON time is the difference of $t_2 - t_1$, and the period is the difference of $t_2 - t_0$; therefore, the duty cycle is:

$$D = \frac{t2-t1}{t2-t0}$$

(Equation 5.50)

Let $t_0 = 0$, $t_1 = 4\mu s$, and $t_2 = 10\mu s$. Substituting these values into *Equation 5.50*, we get:

$$D = \frac{t2-t1}{t2-t0} = \frac{10\mu s - 4\mu s}{10\mu s - 0} = \frac{6\mu s}{10\mu s} = 0.6,$$

or 0.6 * 100 = 60% duty cycle.

(Equation 5.51)

In the above example, the 60% duty cycle is a fixed number, but we want to be able to change the duty cycle from ideal values of 0% to 100%. To achieve this, we can simply change the DC voltage value. To achieve a duty cycle of 10%, we can increase the DC value, as shown in *Figure 5.14a*, and to achieve a duty cycle of 90%, we can lower the DC value, as shown in *Figure 5.14b*.

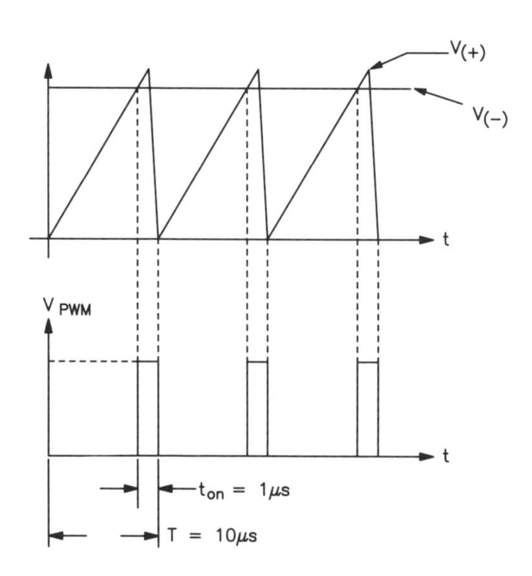

Figure 5.14a

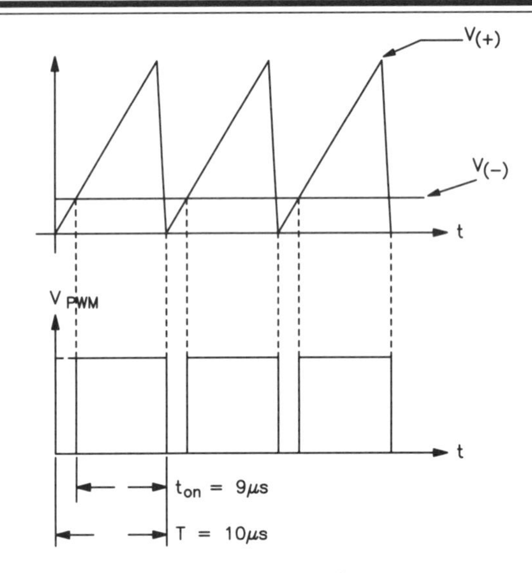

Figure 5.14b

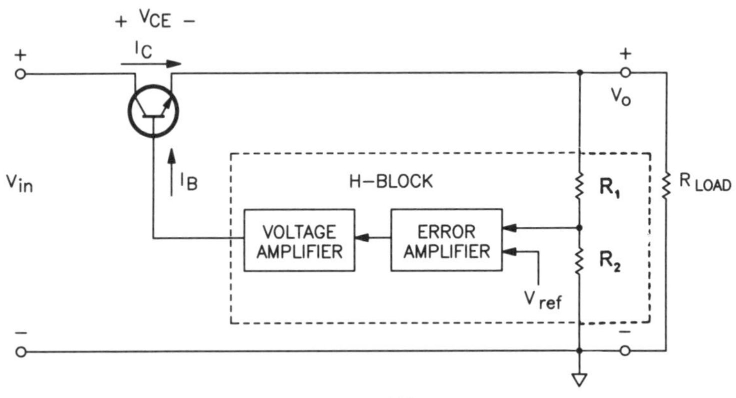

Figure 5.15

5.6 LINEAR CONTROL

To understand the basic concepts of linear control, we will examine the linear regulator of Chapter 3 in detail. It was explained in Chapter 3 that when the output voltage, V_O, goes higher or lower than the nominal value, then the H-block adjusts the parameters to bring the V_O back to its nominal value. We will look at this H-block (see *Figure 5.15*).

R1 and R2 are placed across the V_O, and the voltage V_{R2} is some ratio of V_O. This ratio is derived by the voltage divider circuit. This V_{R2} voltage is compared with V_{ref}. If V_{R2} is higher than V_{ref}, then the difference (V_{R2} - V_{ref})

is amplified by the error amplifier's gain. The output of the error amplifier, which is connected to a base driver circuit, then adjusts the base current to Q, which in turn changes the collector current and V_{CE} drop; thus, the output, V_O, will be adjusted. If V_O is not brought back to its desirable value, then V_{R2} will change again. If the difference is still there, the error amplifier will amplify the difference and adjust the base current to Q. This process will continue until the output is brought to a desirable value.

The amount of time it takes for the output to go from an undesirable value to a desirable one is called a *loop response*. In other words, if the output takes a long time to get to its desired value, then the loop is said to have a slow response time. If, on the other hand, the output is brought back to its desirable value within a very short time, then the loop is said to have a fast response. But if the output does not remain at the desired value and continuously fluctuates around V_O, then the loop is said to be *oscillating*. If the output, V_O, does not remain at a desirable value and keeps increasing or decreasing, or goes to any random value, then the loop is said to be *unstable*.

5.7 DIRECT DUTY CYCLE CONTROL

In the previous chapters, we looked at circuit operations in terms of ON and OFF states. In the PWM section of this chapter, we further defined the ON/OFF states and called it PWM. The pulse-width control or duty cycle control is the subject of this section.

Let's look at how the duty cycle is controlled. We will use the Buck regulator of section 3.3.1 to examine this control technique.

In *Figure 5.16*, the output voltage, V_O, is sensed in a resistor network. V_{R2} is a ratio of V_O, and the ratio is derived from a voltage divider circuit. V_{R2} is compared with V_{ref}, and the difference or error is amplified by the error amplifier. This error is compared with a ramp waveform and the PWM is generated. So, if V_O goes higher than the desired value, V_{R2} will increase, and the DC level at V_{R2} will increase. This will change the duty cycle, which in turn will adjust the output, V_O. Loop response is much faster than in the linear control method. V_{ref} is the reference voltage.

5.8 VOLTAGE FEED FORWARD PWM CONTROL

In this method, the duty cycle is controlled by the input voltage. Unlike the direct duty cycle control method, this method changes the slope of the sawtooth waveform. This technique is used to compensate for any changes in the input voltage; however, the PWM is still controlled by the compensation voltage, as shown in *Figure 5.17*. *Figure 5.18* shows how the PWM waveform changes as a function of the input voltage, V_{in}. In most systems, the output must be maintained at a desirable value under varying input line, which is accomplished by this control technique.

5.9 CURRENT MODE CONTROL

To understand the basic concept of this technique, let's examine the forward converter of Chapter 4 in the current-mode control, because it is most effective in this topology. In the voltage-mode control (direct duty cycle control) technique, we saw that the error voltage was derived from the difference of V_O and V_{ref}. This output (V_O) was then compared with the sawtooth ramp to generate the pulse-width modulated waveform. In this technique, we sensed only the output voltage. This caused V_O to increase or decrease, and the voltage control loop adjusted its parameter to compensate for the change. The whole process may take some time and cause delay in the loop, and may not correct the output voltage within a desired amount of time.

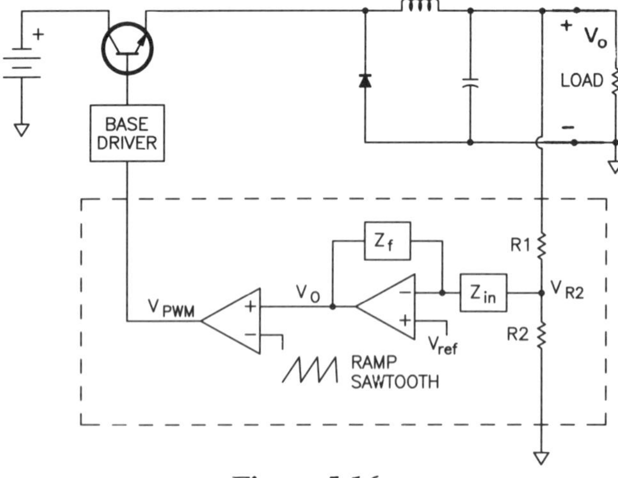

Figure 5.16

Instead of waiting for the voltage to change the duty cycle, the current mode control technique senses the peak current going through the inductor, the primary current, as shown in *Figure 5.19*. This current can be sensed on each pulse, and is known as pulse-by-pulse, or cycle-by-cycle, current sensing. So, in any given pulse, if the amplitude of the current exceeds a certain value, then the loop changes the pulse-width directly instead of waiting for the voltage loop to respond. The voltage loop is called an *outer loop* and the current loop is called an *inner loop*. The inner loop has a fast or instantaneous open-loop response.

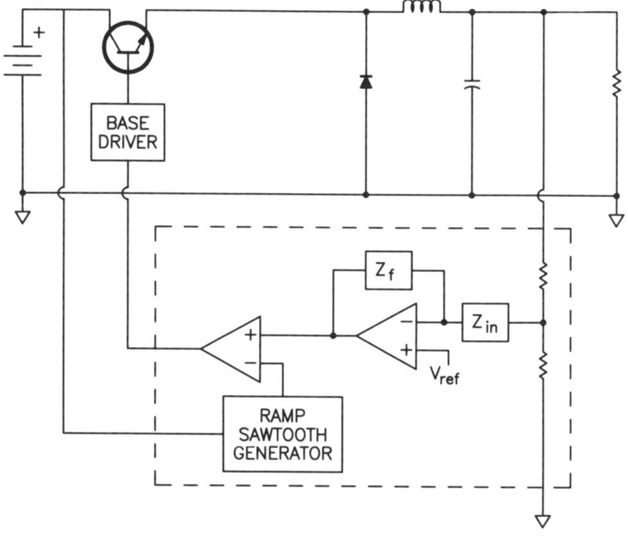

Figure 5.17

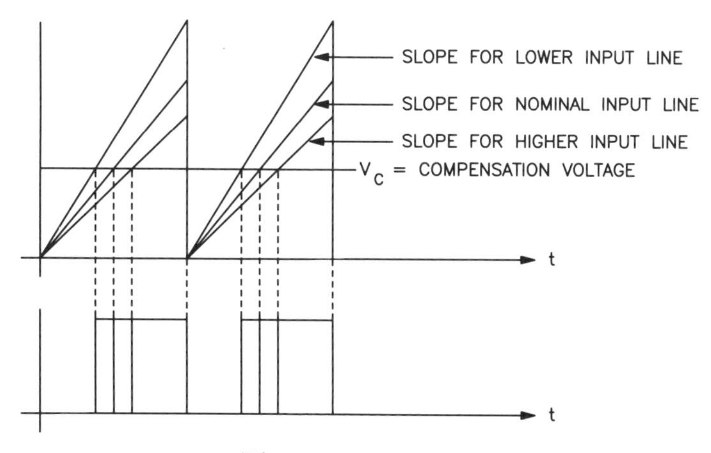

Figure 5.18

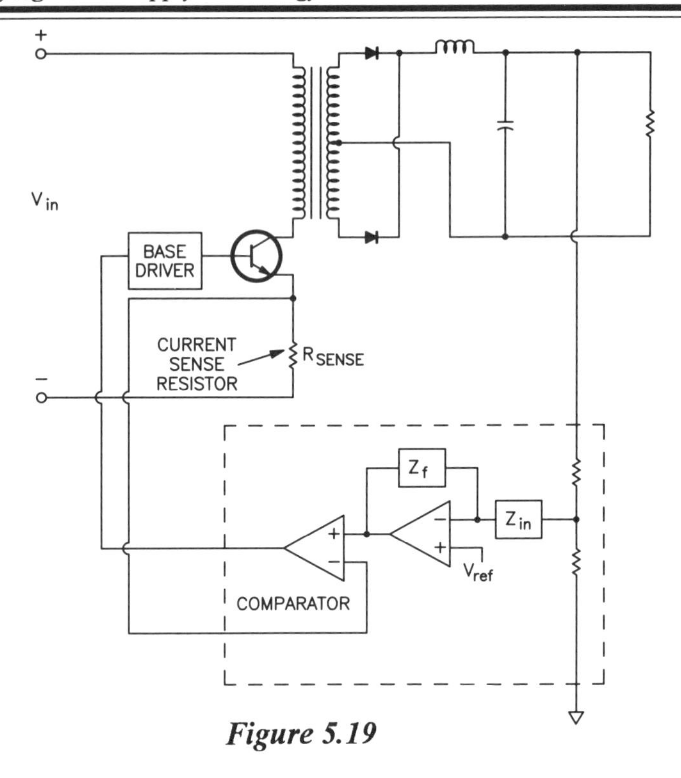

Figure 5.19

The current in the primary circuit is also proportional to the input voltage; therefore, this technique compensates for input line variations.

5.10 ISOLATION AND PROTECTION CIRCUITS

5.10.1 ISOLATION CIRCUITS

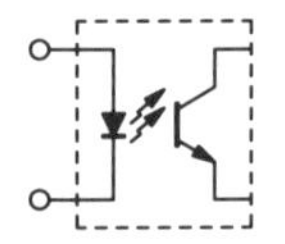

Figure 5.20a

Isolating the H-block for the output can be achieved by various methods, but the most popular one is *optical coupling.* A device called an *opto-isolator* is used for this purpose. The simplest form of an opto-isolator is shown in *Figure 5.20a. Figure 5.20b* shows how the opto-isolator is connected in a circuit. When V_O is present, the current, I_1, will flow into the diode, which is a light-emitting diode. The current through the diode generates light energy and the amount of light energy depends on the current through the diode. This light is transmitted from the diode to the base of the transistor, which is a photo-transistor. When a certain amount of light hits the base of the transistor, I_C will start flowing:

$I_C = h * i_1$, where h is called the current transfer ratio.

I_C flows through the resistor, R2, creating a voltage drop, V_2. This voltage drop is proportional to the input voltage, and I_C is controlled by the light energy which in turn is controlled by the current, I_1. Since R1 is fixed, I_1 is controlled strictly by V_O. We can redraw the circuit of *Figure 5.20b* into a block form showing V_O as the input and V_2 as the output. V_2 becomes the input to the compensation network (see *Figure 5.20b*).

Another isolation circuit may be necessary when we connect the output of the H-block to the input of the G-block. In the switch-mode power supply system, the output from the control section is in PWM form, as opposed to DC form, as was in the previous case. There are two commonly used methods to achieve this isolation. One is the *magnetically-coupled method*, and another is the *optically-coupled method.* A transformer (a magnetically-coupled device) is used to couple the PWM signals from the control section (H-block) to the input circuit of the switching device, as shown in *Figure 5.21*.

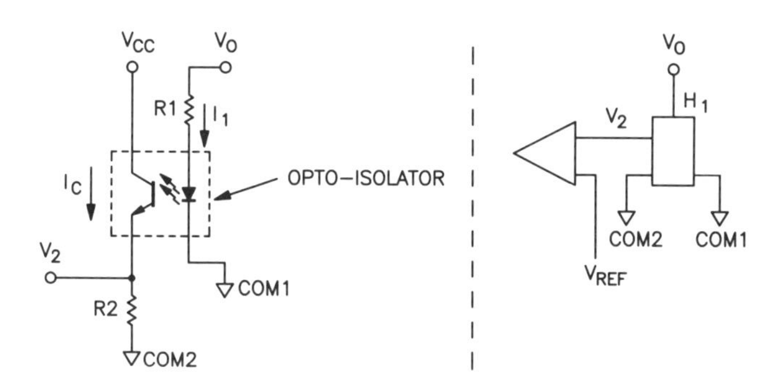

Figure 5.20b

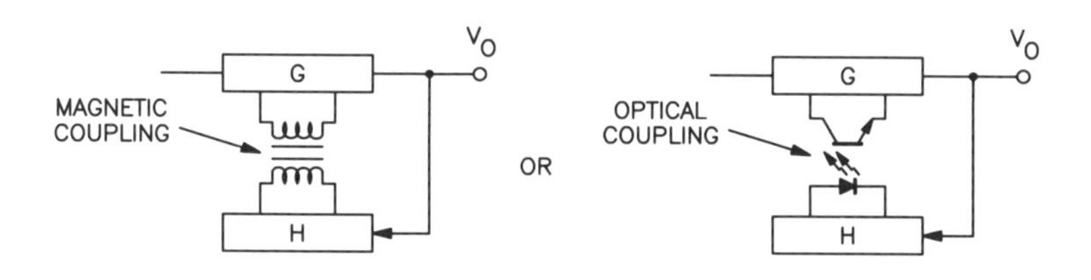

Figure 5.21

5.10.2 PROTECTION CIRCUITS

Overvoltage Circuit

In the event that the control loop cannot maintain the output voltage, the overvoltage circuit either shuts down the system or brings the output voltage to a safe level. Although there are many ways to achieve this function, the basic concept is shown in *Figure 5.22*.

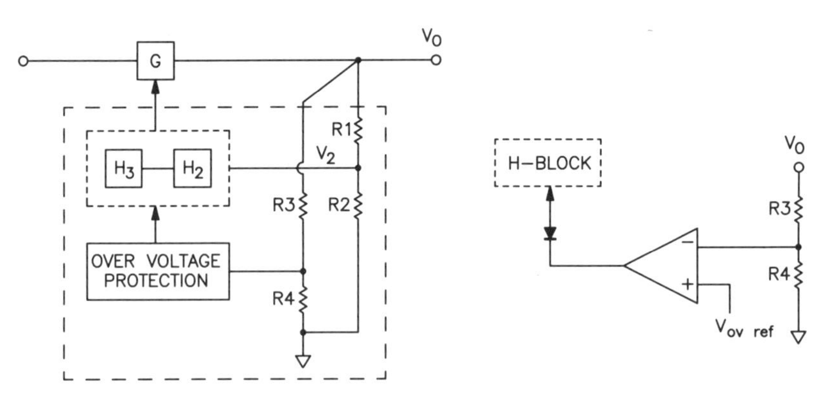

Figure 5.22

If the output voltage exceeds a predetermined value of $V_{OV\ ref}$, then the output of the comparator will become low, forward biasing the diode and disabling the generation of the PWM waveform. For example, let's say that $V_O = +24\ _{DC}$, and we want an overvoltage protection at $+26V_{DC}$. Let's say that $V_{ref} = +5V_{DC}$. So, the voltage divider ratio should be such that at $+26V_{DC}$ output, the $V_{R2} > +5V$.

Undervoltage Circuit

This circuit is similar in principle to the overvoltage circuit. In the overvoltage circuit, the output voltage is sensed. In this circuit, the input voltage is sensed (see *Figure 5.23*).

In *Figure 5.23*, $V_{uv\ ref}$ is connected to the (-) terminal of the comparator and the (+) terminal is connected to the voltage divider circuit. If the voltage divider ratio is less than V_{ref}, then the output will inhibit the generation of the PWM signal; but as soon as $V_{R2} > V_{ref}$, then the output becomes high and the diode becomes reversed-biased. For all practical purposes, we can say that the comparator is disconnected from the PWM generation circuit.

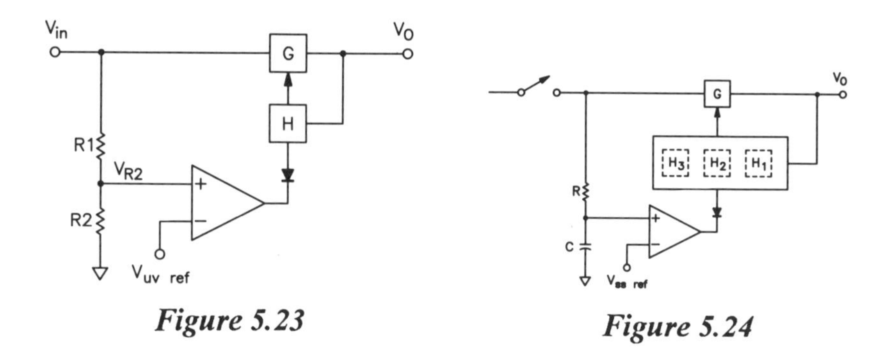

Figure 5.23

Figure 5.24

Soft Start Circuit

Due to the nature of the electronic components, it is necessary to make sure that enough time is allowed for all the components to stabilize. This circuit keeps the PWM generation circuit in disabled mode for some time after application of the power. This is basically accomplished by a time constant of the R-C circuit (see *Figure 5.24*). The time constant is determined by the selection of R and C. Let's say that $V_{ss\ ref} = +5V_{DC}$ and $V_{in} = +30V_{DC}$. When the switch is turned ON, the power is applied to all sections but the PWM circuit. It is disabled by the comparator because the voltage across the capacitor is not charged up to a value higher than $V_{ss\ ref}$.

Overcurrent

Input

If there is a short-circuit condition in the system, it will draw an excessive amount of current from the input source. A fuse or a circuit breaker is normally used to protect the source from failing. Both devices open the path to the system, stopping the current flow.

Output

When the output current exceeds the specified maximum current, then the current sensing circuit either limits the current at a maximum value or the current folds back to a lower current limit.

CHAPTER 6
POWER SUPPLY SYSTEM SPECIFICATIONS AND APPLICATIONS

6.1 INTRODUCTION

A complete power supply system consists of various sections, as described in the previous chapters. When we are dealing with individual sections, we are only talking about the characteristics of each section; but when we are dealing with the entire power supply system, we must deal with the effects of combining all of these sections together. The choice of selecting different functional blocks depends mainly upon the specifications that the system is required to meet. The specifications may be generated either by the customer or by the designers. In any case, the requirements tell a lot about the type of environment in which the power supply will be used, so a good understanding of specifications is very important. This chapter will focus on defining specifications, and we will see various configurations in the applications.

6.2 SPECIFICATIONS

6.2.1 INPUT

Input Voltage: The nominal voltage supplied by the source. This voltage is usually given in a nominal value, +/- some tolerance. If the input voltage is $120V_{AC}$(rms), tolerance may vary +/- 10%, or V_{in} (minimum) = 120 - 0.1(120) = $102V_{rms}$, and V_{in} (maximum) = 120 + 0.1(120) = $142V_{rms}$. The input source can be in either AC or DC form. If the input is AC, then it can be either one phase or more. Input voltage requirements depend on where the power supply system will be used. *Table 6.1* shows typical input voltages for various parts of the world.

Some manufacturers design power supplies to accept universal input, which covers all of the nominal voltage ranges around the world. Universal input voltage range can be from $85V_{rms}$ - $265V_{rms}$.

COUNTRY	VOLTAGE (V_{rms})
Australia	240
Brazil	127
Canada	120
Denmark	220
Finland	220
Germany	220
Ireland	220
Italy	220
Japan	100
Mexico	127
Netherlands	220
Sweden	220
Switzerland	220
Taiwan	110
U.K.	240
U.S.A.	120

Table 6.1

Input Frequency: If an input source is of AC type, then the frequency is specified with it. The input frequency can vary from 47Hz - 440Hz. In the United States, the common household voltage frequency is 60Hz. The frequency also has a tolerance. Nominal frequencies for various parts of the world are shown in *Table 6.2*.

Input Voltage Transients: If power supplies are operated in a noisy environment, then the input voltage (besides the nominal V_{in} - max) may see high voltage spikes of some duration. This specification is critical because the components of the power supply must be able to withstand it.

Inrush Current: The surge of current flowing into the power supply when it is first turned ON. This current depends on the input voltage. For example, $120V_{AC}$ input will have a lower surge current than $240V_{AC}$.

6.2.2 OUTPUT

Voltage Level: In the fixed output power supply, the voltage is given in nominal value, +/- some tolerance. If the power supply has an adjustable output, then this voltage is specified as a *range*. For example, 0 - $+24V_{DC}$, or in many cases, the voltage range may be only a small range; $+20V_{DC}$ to $+28V_{DC}$, etc. In the multiple output power supply, the outputs can be of different polarities; i.e. output #1 could be $+5V_{DC}$, output #2 could be $-5V_{DC}$, etc.

Number of Outputs: The power supply can be a single output or a multiple output.

Current: In voltage regulated power supplies, current is specified as a maximum output current from the power supply.

Line Regulation: The amount of variation from the nominal output voltage, when the input line changes from the lowest point of the input voltage range to the highest point. For example, if the input line is specified as 120V +/- 10%, then the output voltage level is at 102V (120 - 10% of 120), and the output level is at 142V (120 + 10% of 120). This deviation in voltages, with respect to the nominal output voltage, is usually specified in a percentage; i.e. +/- 0.1% maximum, etc. The load is assumed to be at a constant value.

COUNTRY	FREQUENCY (Hz)
Australia	50
Brazil	60
Canada	60
Denmark	50
Finland	50
Germany	50
Ireland	50
Italy	50
Japan	50/60
Mexico	60
Netherlands	50
Sweden	50
Switzerland	50
Taiwan	60
U.K.	50
U.S.A.	60

Table 6.2

Load Regulation: The amount of variation from the nominal output voltage, when the output load is changed from no-load to full-load. When there is a no-load connected, then the output voltage level is different from when a full-load is connected. This deviation is usually given in a percentage. The line is assumed to be at a constant value.

Overshoot/Undershoot: The amount of voltage deviation (higher or lower) at turn ON or turn OFF of the power supply, or at the change of load.

Transient Response Time: The amount of time it takes for the output to return to its nominal voltage range, when the output load changes from one level to another.

Temperature Coefficient: The amount of output voltage changes when the ambient temperature changes. It is usually given as a slope; i.e. 0.002%/C. For example, if $V_O = 24V_{DC}$ at a room temperature of 20^0 C, then the temperature changes to 70^0 C. The total output voltage will change as follows:

$$\frac{0.00002}{C} * (70C-20C) * 24V = \frac{0.00002}{C} * 50C = (0.00002) * 50 * 24V$$
$$= 0.024V \text{ or } 24mV$$

Therefore, the total change in the output voltage due to the temperature variation is 24mV.

Holdup Time: The amount of time the output voltage remains regulated after the input source is disconnected. This parameter is normally measured at a full load, either nominal or minimum line voltage.

Turn-On Time: The amount of time it takes for the output voltage to reach the nominal value after the input power is applied.

Drift (or Stability): The output voltage drifts to a different value after the temperatures of the components have stabilized. It is normally given as percent change of the output over a certain period of time; i.e. the warm-up period can be specified as 30 minutes, etc.

Ripple and Noise: The amount of AC voltage on the output. AC voltage consists of a noise from the power supply and the random noise. This is specified either in peak-to-peak or *rms* value over some bandwidth.

Short Circuit (Duration): The amount of time the output can remain in short circuit condition without causing any damage to the power supply system. There are current limit specifications, such as maximum current and/or foldback current.

6.3 APPLICATIONS

As mentioned earlier, power supplies are made up of various functional blocks, but power supplies are also connected to the outside world. We will see how they become connected and discuss the various configurations.

Power supplies may be connected in series or parallel, etc. **It should be noted that the following circuits are shown only to explain the concepts, and the author does not make any suggestions as to how some configurations should be made. A great deal of attention must be given to other parts of the system before making any connections.**

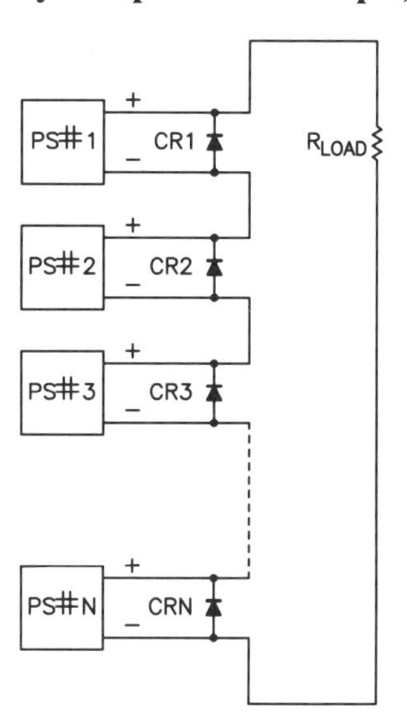

Figure 6.1a

6.3.1 SERIES CONNECTIONS

In certain applications, it may be necessary to connect two or more power supplies in series, as shown in *Figure 6.1a*. For understanding the basic concepts, let's examine a configuration with only two power supplies, PS #1 and PS #2 (*Figure 6.1b*). This type of configuration gives the capabilities of dual polarities. PS #1 can provide the positive output voltage with respect to the common terminal, while PS #2 can provide the negative output voltage. This configuration can also give an output voltage sum of $V_{01} + V_{02}$. CR1 and CR2 are the protection diodes. In the event that the load is in short-circuit condition, these diodes will prevent the output of the power supply from being exposed to the reverse voltage of the other supply.

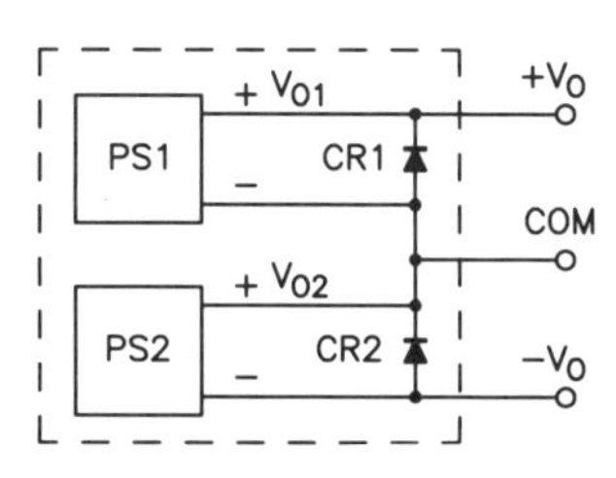

Figure 6.1b

6.3.2 PARALLEL CONNECTIONS

Some power supplies are specifically designed to operate in parallel mode. The basic idea is to increase the current capabilities, thereby increasing the power delivering capability of the total power supply system.

The system consists of two or more power supplies in parallel, as shown in *Figure 6.2a.* There are several variations to this basic concept. First we will examine the circuit of *Figure 6.2a* and discuss the limitations of it, then move on to other configurations.

Let's examine the configuration with two power supplies, PS #1 and PS #2 (*Figure 6.2b*). In this configuration, both power supplies have exactly the same output voltages. Ideally, if they are connected in parallel, and both are connected to the same load, then they will deliver the same amount of current. If the load requires 6A, then 3A from each power supply is provided. This is called *load sharing*; however, in real life, it is very difficult to find two power supplies that will maintain exactly the same output voltages at all times.

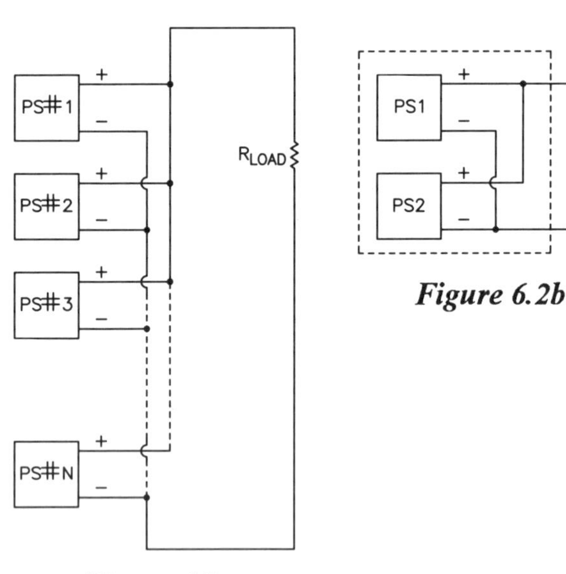

Figure 6.2b

Figure 6.2a

Suppose that two power supplies are connected in parallel, and the output voltage of one of them changes slightly. Then the power supply with the higher output voltage will supply more current, and the one with the lower voltage will supply less current; but notice that the total power supplied is still the same.

Now suppose that the power supply with the higher voltage keeps going higher, delivering more current. At some point it will deliver the full power. Normally, power supplies connected in this mode do not have full power capability; the power capability of each power supply is usually half if only two power supplies are connected. If three are connected, then it is 1/3 (because the total power requirement is now 3 times, etc.).

If only one power supply is trying to deliver power that is twice its capacity, it will exceed the maximum current limit and go into overcurrent mode, and it will either remain there or go into foldback mode. In the overcurrent mode, the output voltage will go lower. In this case, the other power supply will start delivering more current, and after some time, it may go into the overcurrent limit. This could damage the power supplies.

There are several techniques for overcoming this problem, and one of them is shown in circuit *Figure 6.3*. It uses a resistor in series with an output terminal, which compensates for a slight change in the output voltage.

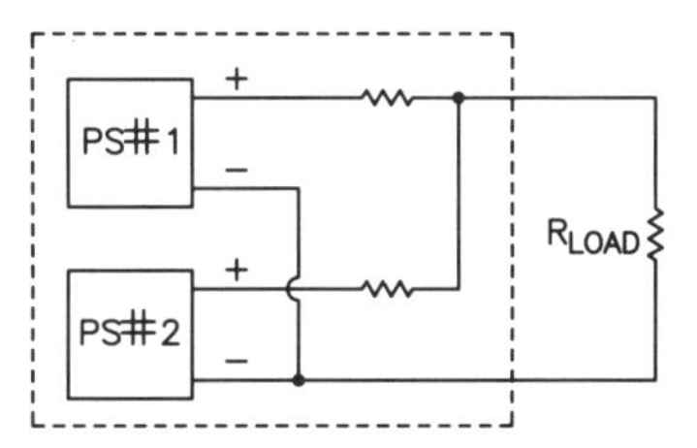

Figure 6.3

6.3.3 MASTER/SLAVE CONNECTIONS

Series: In the previous configuration of series connection, the output voltages of each power supply were controlled independently by the corresponding control loop. In this configuration (see *Figure 6.4*), there is only one loop that controls the output of all the power supplies. The power supply that controls the output of all the other power supplies is called a Master, while the others are called slaves. The concept is simple to understand; the Master commands and the slaves follow.

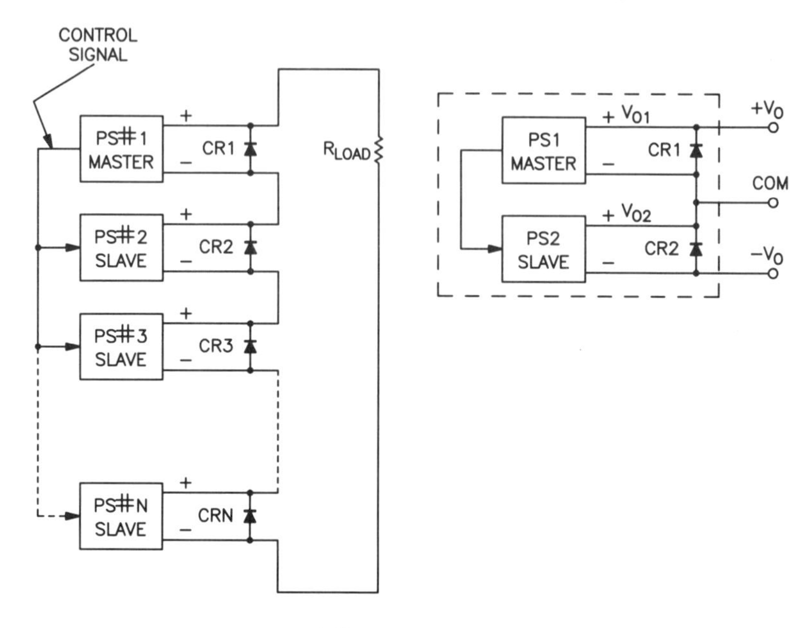

Figure 6.4

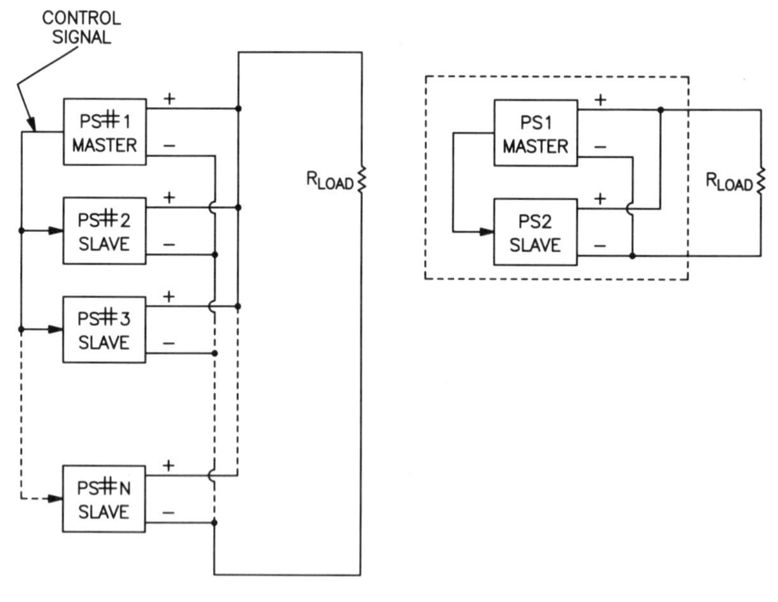

Figure 6.5

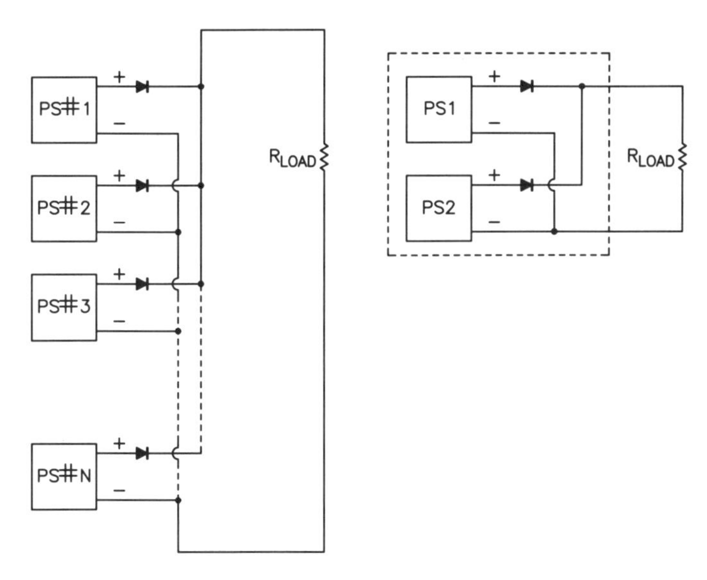

Figure 6.6

Parallel: Similar to the series Master/slave configuration, the output voltages of each power supply connected in parallel are controlled by the Master. The configuration is shown in *Figure 6.5*.

Redundant Configuration (Parallel): In this configuration, the outputs of each power supply are still connected in parallel, but through the diodes, as shown in *Figure 6.6*. In the previous configuration, the power capabilities of each power supply were not necessarily a full power

There are two commonly used methods:

1) The full power is supplied by only one power supply, while the others don't supply any power.
2) All power supplies deliver equal power to the load, and in case one fails, then the others will have enough capacitance to take over.

6.3.4 VOLTAGE SENSING

There are two commonly used methods for sensing the output voltage, V_O. Recall that this voltage is the input voltage to the H-block, and it is connected to the voltage divider. Now we will examine how this is sensed.

The connection can be called *local sense* or *remote sense.* In local sense configurations, the sense connection is done at the output terminals of the power supply. In the remote sensing configuration, this connection is made at the load. One might ask, what is the difference? They are both in parallel and the voltages across the parallel connections are equal. True, but these parallel connections are made using physical cables.

In DC applications, the cables have resistance, and when the output current goes from the power supply to the load, the voltage (I * R) is dropped across the resistance of the cable. If the current is high, then this voltage drop may be significant, and the voltage across the load is less than what is required. Also, the power (i^2 * R) is dissipated in these cables. So, the power supply loop thinks it is regulating an output voltage, which is true, but not across the load (at the output terminals).

To overcome this problem, if we sense the output voltage directly at the load, then the loop can compensate for the voltage drop across the cables. So, the sense connections are made at the terminals of the load instead of at the power supply. Since the load is located some distance away or at a remote location from the power supply, this configuration is called a *remote sense configuration.*

Because the power supply is compensating for the voltage drop across R, it is generating a higher voltage than in the previous case. Since the load current is the same, the output power will be higher because "higher" voltage * the same load current = "higher" power.

6.3.5 REMOTE PROGRAMMING

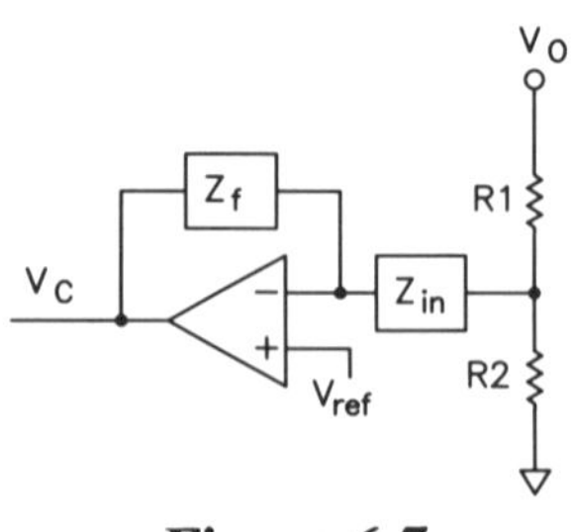

Figure 6.7

In the variable power supplies, the output voltages are adjusted by a potentiometer, a power supply, an externally connected potentiometer, or an externally connected voltage source. But in all cases, a common method of controlling V_{ref} is used, as in the compensation network shown in *Figure 6.7.* The figure shows that V_{ref} is connected at the (+) terminal of the op-amp. Let's examine these connections.

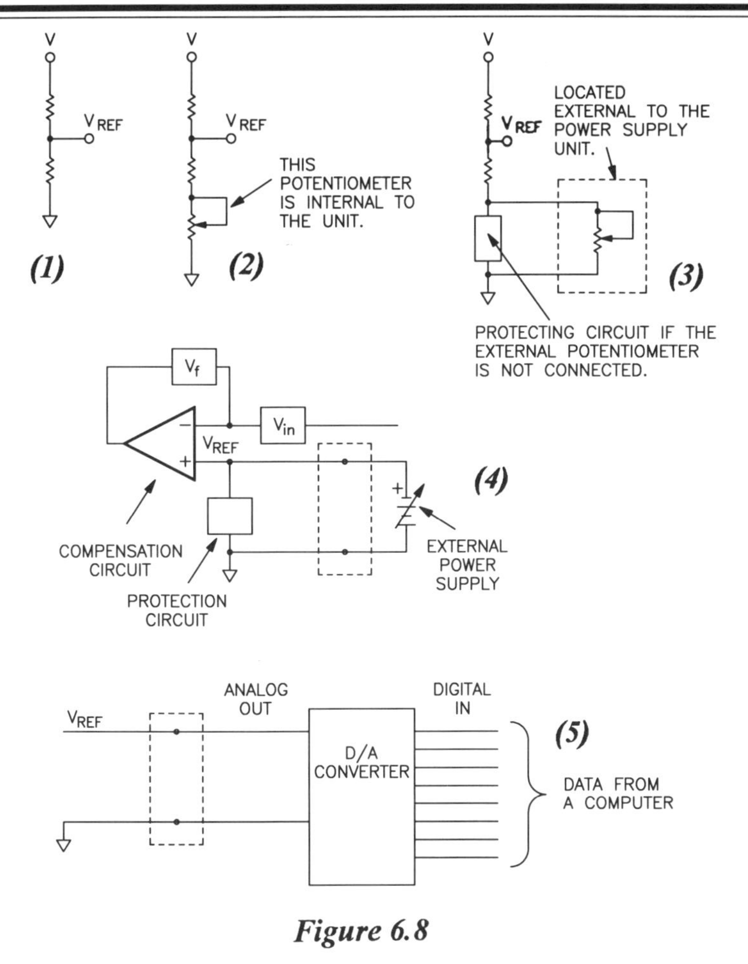

Figure 6.8

The external supply can be either a variable voltage or a computer D/A converter from a computer system (see *Figure 6.8*). Configuration #1 is for the fixed output and #2 is for the local adjustments. Configurations #3, #4 and #5 are called remote programmingm.

6.4 REGULATORY AGENCIES

Since power supplies carry various levels of voltage, it is the manufacturers' responsibility to make sure that these voltages are isolated from the outside. Proper isolation must be provided, but "proper isolation" could mean different things to different manufacturers. To avoid such ambiguities, standards were developed by regulatory agencies. These types of agencies exist both at a national level and at an international level.

Safety Agencies: *International*

IEC - International Electro-technical Commission.
CEE - International Commission on Rules for the Approval of Electrical Equipment.

Safety Agencies: *National*

VDE - Verband Deutscher Electrotechniker (West Germany).
UL - Underwriter's Laboratories (United States).
CSA - Canadian Standards Association (Canada).
BS - British Standards (United Kingdom).

The regulating agency for the Electromagnetic Interference (EMI) is the FCC (Federal Communications Commission) in the United States and the VDE in Europe.

GLOSSARY

The words in this section are not necessarily used in the previous chapters, but they are encountered in the power supply industry.

Ambient Temperature: The temperature of the surrounding environment.

Bandwidth: The frequency at which the gain has fallen off by 3_{dB}.

Bleeder Resistor: A resistor that is connected across a capacitor, usually across an input filter capacitor. The purpose of this resistor is to discharge the capacitor when the input voltage is removed. If this resistor is not connected across the capacitor, then the capacitor may not discharge fully and some voltage may be present long after the input power is removed.

Brownout: When the line voltage is reduced by about 15 - 20% from its nominal value.

Brownout Protection: The ability of a power supply to perform normally during the brownout period.

Burn In: A period during which a newly assembled power supply unit is tested. This testing is usually done at a nominal line and a full-load.

Bus: The main conductor that delivers the power from the power source to the load.

Common Mode Noise: A noise that is common to both input lines or output lines with respect to a reference point, such as a chassis ground.

Constant Current: If a system delivers a fixed amount of current under a varying load, then the system is said to be operating in constant current mode.

Constant Current Limiting Circuit: A circuit that limits the output current to a maximum value when an overload condition exists.

Constant Voltage Power Supply: A power supply system that maintains a fixed value of output voltage under varying conditions of input line voltage, output load, or ambient temperature.

Cooling: Removal of heat from a power supply system.

Cross Regulation: In a power supply system with multiple outputs, the voltage of one output may change due to the variation in load of other outputs. The ratio of this change to the nominal output voltage is called the *cross regulation*.

Crowbar: A method used to short out the power supply output to ground in an overvoltage condition.

CSA: Canadian Standards Association. It is a national regulatory agency in Canada which defines the safety standards for systems such as power supplies.

Current Limiting Circuit: A circuit that limits the output current to a predetermined value.

Cycle-By-Cycle Current Limiting Circuit: This circuit monitors each current pulse in a PWM cycle. If the current in any cycle exceeds a predetermined value, then this circuit will limit the amount of current delivered by the system.

Derating: The power rating of a power supply system is reduced to less then the rated value. The rated value may be given at a nominal temperature, but as the temperature increases, the rated power that the power supply delivers is reduced.

Differential-Mode Noise: The component of noise that is measured between two lines. It is measured between the DC output and DC output return.

Dynamic Load: If a load that is connected across the output terminals of a power supply changes rapidly to a different value, it is called a dynamic load.

Efficiency: A ratio of total output power to input power that is given in percentage. In a multiple-output system, the total power is the sum of power delivered by individual output. It is normally specified at a nominal line and a full-load of each output.

EMI: Electromagnetic Interference. Undesirable high-frequency noise generated due to the switching actions of electronic devices within the power supply system.

ESR: Equivalent Series Resistance. A real capacitor can be modeled using R, C, and L components.

FCC: Federal Communications Commission.

Filter: A network that passes desirable frequencies and attenuate unwanted frequencies.

Foldback Current Limiting Circuit: A protection circuit that reduces the current as the output load increases and limits to a lower than maximum current limit.

Hi-Pot: High Potential. Used to test the dielectric strength of an insulator.

Holdup Time: The time in which the power supply system can deliver the regulated output after the input power is removed.

IEC: International Electrotechnical Commission.

Inrush Current: The peak current flow into the power supply when the AC power is first applied.

Inrush Current Limiting: The circuit that limits the inrush current.

Inverter: A device that converts DC power to AC.

Isolation: An electrical separation between the input and output of a power supply system, usually achieved by a transformer.

Line Regulation: The maximum percentage change in the output voltage when the input line changes from the high line to low line, and vice versa.

Load Regulation: The maximum percentage change in the output voltage when the output load changes from no-load to full-load.

Master: In the Master-slave operation, this is the unit the controls the output of the other units.

MTBF: Mean Time Between Failures. It is a measure of reliability. A system with a higher MTBF is considered to be more reliable than a system with a lower MTBF.

Multiple Output Supply: A power supply with more than one output.

Off-Line Switcher: The input section of a switching power supply that is directly connected across the input line.

Operating Temperature: The temperature range in which the power supply system will operate within the specifications.

PARD: Periodic And Random Deviation.

Power Factor: A ratio of real power to apparent power.

Pulse-Width Modulation: A technique used in switching power supplies where the pulse width of a switching wave is modulated to control the output of a power supply.

Static Load: If a load that is connected across the output terminals of a power supply remains constant, it is called a static load.

Switching Frequency: The frequency at which is switching devices operate in a switching power supply.

UL: Underwriters' Laboratories. A safety regulatory agency in the United States.

UPS: Uninterruptible Power Supply.

VDE: Verband Deutscher Elektrotecniker. In Europe, its functions are similar to those of UL in the United States.

BIBLIOGRAPHY

Alphabetical by Author

P.R.K. Chetty, *Switch-mode Power Supply Design*, TAB Professional and Reference Books, Blue Ridge Summit, PA, 1986.

George Chryssis, *High-frequency Switching Power Supplies*, McGraw-Hill Book Company, Blue Ridge Summit, PA, 1984.

Computer Products, *Power Supply Engineering Handbook* (1990), Computer Products, Fremont, CA.

John J. D'Azzo & Constantine H. Houpis, *Feedback Control System Analysis and Synthesis,* McGraw-Hill Book Company, Blue Ridge Summit, PA, 1966.

Chris Everett, "High-frequency Off-line Switching Power Supplies," *EDN*, v.31, pp. 130-135, April 17, 1986.

Paul R. Gray, *Analysis and Design of Analog Integrated Circuits, Second Edition,* John Wiley & Sons, New York, NY, 1984.

Thomas L. Floyd, *Principles of Electric Circuits,* Charles E. Merrill Publishing Co., 1981.

Eugene R. Hnatek, *Design of Solid-State Power Supplies,* Van Nostrand Reinhold Company, New York, NY, 1981.

William H. Hyat, Jr. and Jack E. Kemmerly, *Engineering Circuit Analysis (Third Edition)*, McGraw Hill Book Company, Blue Ridge Summit, PA, 1978.

Robert G. Irvine, *Operational Amplifier Characteristics and Applications*, Prentice Hall, Inc., Englewood Cliffs, NJ, 1981.

R. Itoh & K. Ishizaka, "Single-phase Sinusoidal Rectifier with Step-up/ Step-down Characteristics," *IEE Proceedings. Part B, Electric Power Applications* v.138 pp. 338-44 , November 1991.

Edward C. Jordan, *Reference Data for Engineers: Radio, Electronics, Computer, and Communications (seventh edition),* Howard W. Sams & Company, Indianapolis, IN, 1989.

Kepco Power Supplies, *Applications Handbook & Catalog (146-1457),* Kepco, Inc., Flushing, NY.

Kepco Power Supplies, *Kepco Instrumentation Power Supplies for System and Bench (146-1716),* Kepco, Inc., Flushing, NY.

Craig Maier, "Taking the Mystery Out of Switching Power-Supply Noise," *Electronic Design* v.39, pp. 83-92, September 26, 1991.

John Mayer, "Switching Power Supplies Cut Size, Improve Reliability," *Computer Design,* v.26, pp.96-100, June 1, 1987.

R.D. Middlebrook and Slobodan C'uk, *Advances in Switched-Mode Power Conversion (VOL. I , II and III),* Teslaco, 1983.

R.D. Middlebrook, "Small-signal Modeling of Pulse-width Modulated Switched-mode Power Converters," *Proceedings of the IEEE,* v.76, pp. 343-354, April 1988.

Motorola, *Switchmode: A Designer's Guide for Switching Power Supply Circuits and Components - 1992 (SG79/D, rev.4),* Motorola, Inc., 1992.

Abraham I. Pressman, *Switching and Linear Power Supply, Power Converter Design,* Hayden Book Company, Inc., Rochelle Park , NJ, 1985.

Warren Schultz, "Power Transistor Safe Operating Area—Special Consideration for Switching Power Supplies," Application Note AN-875, Motorola Semiconductor Products, Inc.

Warren Schultz, *Understanding Power Transistor Dynamic Behavior—dv/dt Effects on Switching and RBSOA*, Application Note AN-873, Motorola Semiconductor Products, Inc.

Jeffrey D. Shepard, *Power Supplies,* Reston Publishing Company, Inc., Reston, VA, 1984.

Robert L. Steigerwald, "A Comparison of Half-Bridge Resonant Converter Topologies," *IEEE Trans. Power Electronics*, vol. 3, no.2, pp. 174-182, April 1988.

Dan Strassberg, " Power-Factor Corrected Switching Power Supplies," *EDN* v.36, pp. 90-96, April 11, 1991.

Ralph E. Tartar, *Principles of Solid-State Power Conversion,* Howard W. Sams & Co., Indianapolis, IN, 1985.

Unitrode Power Supply Design Seminar Handbook, SEM-500 (1986), SEM-700 (1990), SEM-800 (1991), Unitrode Corporation, Lexington, MA.

INDEX

C

D

E

F

G

H

I

K

L

M

From the Engineers of
Howard W. Sams & Company
SEMICONDUCTOR
CROSS REFERENCE BOOK
PROMPT

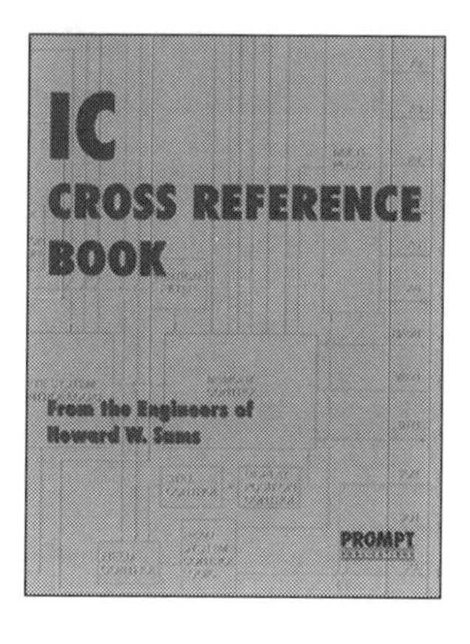
IC
CROSS REFERENCE
BOOK
From the Engineers of
Howard W. Sams
PROMPT

More Technical Titles from PROMPT® Publications—

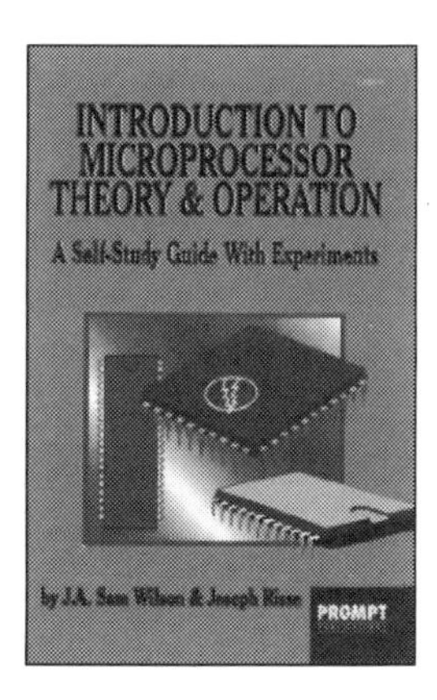

Introduction to Microprocessor Theory & Operation

A Self-Study Guide With Experiments

by Wilson & Risse

This book takes you into the heart of computerized equipment and reveals how microprocessors work. By covering digital circuits in addition to microprocessors and providing self-tests and experiments, this book makes it easy to learn microprocessor systems.

$16.95

Paper/212 pp./6 x 9"

ISBN #: 0-7906-1064-7

Schematic Diagrams

The Basics of Interpretation and Use

by J. Richard Johnson

Step by step, *Schematic Diagrams* shows you how to recognize schematic symbols and their uses and functions in diagrams. You will also learn how to interpret diagrams so you can design, maintain, and repair electronics equipment.

$16.95

Paper/208 pp./6 x 9"

ISBN #: 0-7906-1059-0

Call 800-428-7267 TODAY for the name of your nearest PROMPT® Publications distributor. Be sure to ask for your free PROMPT® catalog!

PROMPT® Publications is your best source for informative books in the technical field.

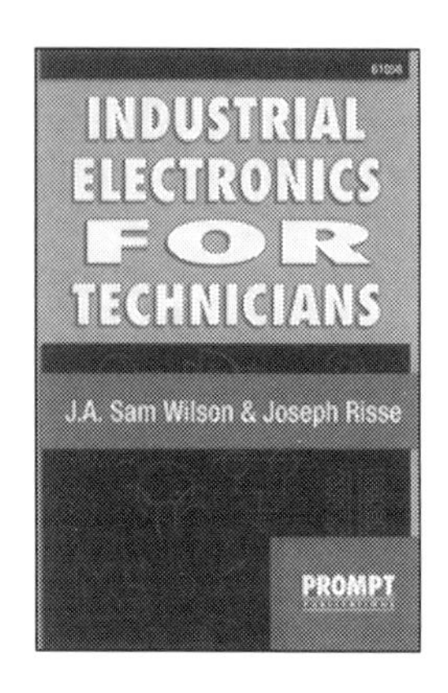

Electronic Control Projects for the Hobbyist and Technician

by Henry C. Smith & Craig B. Foster

Each project in *Electronic Control Projects* involves the reader in the actual synthesis of a circuit. A complete schematic is provided for each circuit, along with a detailed description of how it works, component functions, and troubleshooting guidelines.

$16.95

Paper/168 pp./6 x 9"

ISBN #: 0-7906-1044-2

Industrial Electronics for Technicians

by Wilson & Risse

Increase your career potential by becoming a certified technician in the fast-growing field of industrial electronics. This book provides an effective overview of the topics covered in the Industrial Electronics CET test, and is also a valuable reference on industrial electronics in general.

$16.95

Paper/340 pp./6 x 9"

ISBN#: 0-7906-1058-2

Call 800-428-7267 TODAY for the name of your nearest PROMPT® Publications distributor. Be sure to ask for your free PROMPT® catalog!